Todor Dimitrov Ganchev
Computational Bioacoustics

Speech Technology and Text Mining in Medicine and Health Care

Edited by
Amy Neustein

Volume 4

Published in the Series

Neustein (Ed.), *Text Mining of Web-Based Medical Content,* 2014
ISBN 978-1-61451-541-8, e-ISBN 978-1-61451-390-2,
e-ISBN (EPUB) 978-1-61451-976-8, Set-ISBN 978-1-61451-391-9

Neustein (Ed.), *Speech and Automata in Healthcare,* 2014
ISBN 978-1-61451-709-2, e-ISBN 978-1-61451-515-9,
e-ISBN (EPUB) 978-1-61451-9607, Set-ISBN 978-1-61451-516-6

Beals et al., *Speech and Language Technology for Language Disorders,* 2016
ISBN 978-1-61451-758-0, e-ISBN 978-1-61451-645-3,
e-ISBN (EPUB) 978-1-61451-925-6

Todor Dimitrov Ganchev

Computational Bioacoustics

—

Biodiversity Monitoring and Assessment

DE GRUYTER

Author
Prof. Todor Ganchev
Department of Computer Science & Engineering
Technical University of Varna,
9010 Varna
Bulgaria
tganchev@tu-varna.bg

ISBN 978-1-61451-729-0
e-ISBN (PDF) 978-1-61451-631-6
e-ISBN (EPUB) 978-1-61451-966-9
Set-ISBN 978-1-61451-632-3
ISSN 2329-5198

Library of Congress Cataloging-in-Publication Data
A CIP catalog record for this book has been applied for at the Library of Congress.

Bibliographic information published by the Deutsche Nationalbibliothek
The Deutsche Nationalbibliothek lists this publication in the Deutsche Nationalbibliografie;
detailed bibliographic data are available on the Internet at http://dnb.dnb.de.

© 2017 Walter de Gruyter Inc., Boston/Berlin
Typesetting: Integra Software Services Pvt. Ltd.
Printing and binding: CPI Books GmbH, Leck
Cover image: MEHAU KULYK/SCIENCE
PHOTO LIBRARY / Agentur Focus
♾ Printed on acid-free paper
Printed in Germany

www.degruyter.com

Abstract

This book offers an overview of some recent advances in the computational bioacoustic methods and technology. In the focus of discussion is the pursuit of scalability, which would facilitate real-world applications of different scope and purpose, such as wildlife monitoring, biodiversity assessment, pest population control, and monitoring the spread of disease-transmitting mosquitoes. The various tasks of computational bioacoustics are described and a wide range of audio parameterization and recognition tasks related to the automated recognition of species and sound events are discussed. Many of the computational bioacoustic methods were originally developed for the needs of speech, audio, or image processing, and afterwards were adapted to the requirements of automated acoustic recognition of species, or were elaborated further to address the challenges of real-world operation in 24/7 mode. The interested reader is encouraged to follow the numerous references and links to web resources for further information and insights. This book is addressed to software engineers, IT experts, computer science researchers, bioacousticians, and other practitioners concerned with the creation of new tools and services, aimed at enhancing the technological support to computational bioacoustic applications.

Preface

This book was conceived with the idea to provide an overview of basic methods in *Computational Bioacoustics* and to highlight some important recent developments. Specifically, we illustrate the applicability of a wide range of audio parameterization and pattern recognition methods on tasks aiming at the automated acoustic recognition of species and sound events. Highlighting the technological aspects, we discuss the specifics of various computational bioacoustic tasks that are essential to achieving scalability of biodiversity assessment and monitoring studies. In addition, we provide implementation details and examples on the application of state-of-the-art signal processing and machine learning methods, which facilitate coping with key challenges such as robust operation in real-world conditions. In this regard, along with the traditional audio analysis and sound events recognition methods, we describe some image processing techniques applied on the audio spectrogram seen as an image and current advances in the area of multi-label multi-instance machine learning. Finally, we analyse the technological innovation brought by some recent application-oriented projects, among which ARBIMON, SABIOD, ENTOMATIC, and REMOSIS that made significant contributions to the automation of biodiversity monitoring studies, pest population control in agriculture, and monitoring the spatial distribution of mosquito populations that pose a major threat to human health. These projects are considered the present-day success stories that illustrate well the perspectives opened due to the advances in computational bioacoustics.

Another, and even stronger, stimulus that motivated the emergence of this book was the need for popularizing computational bioacoustic methods and applications among young researchers and engineers, and especially those who already followed an introductory course on speech, audio, or image processing technology. We will be glad if this book would hopefully attract, inspire, or encourage talented researchers to join the growing computational bioacoustic community and pursue a career that will contribute to the efforts to protect and preserve biodiversity. The author is hopeful that the improved awareness of governmental institutions and policymakers about the importance of automated biodiversity monitoring and the opportunities that technological support offers to wildlife protection, environment monitoring, and pest populations automated control applications will stimulate the growth of job openings for technicians, engineers, and data scientists who specialize in the area of computational bioacoustics.

We hope that the interested reader who will go through all book chapters will reap the benefits of an improved understanding of the role, use, and applications of computational bioacoustics, as well as its importance in the global efforts for slowing down the loss of biodiversity. Broadening our knowledge about tasks, methods, and the inherent limitations of technology would allow appropriate practical use of tools and services, as well as the correct interpretation of result obtained through these. The correct comprehension of the potential for automation will establish the ground for future advances in information retrieval from soundscape recordings, a prerequisite for improved technological support to wildlife monitoring and conservation actions.

This volume is not intended to serve as an academic textbook. However, it could be used as a supplementary reference in undergraduate or graduate courses that embrace topics on bioacoustics, ecoacoustics, or general topics associated with audio and image processing applications. Master and PhD students will learn various facets of audio parameterization and acquaint with some well-conceived applications of contemporary pattern recognition methods. Engineers and developers of applications will advance their understanding of methods behind some less-known technological workarounds and trade-offs in computational bioacoustics.

The manuscript preparation coincided with traumatic period in the author's personal life, and the completion of this book has become a challenge. The author would like to express gratitude and acknowledge the understanding and empathetic support by the book series editor, Dr Amy Neustein, and the publisher (Dr Alex Greene and his team), who made possible the emergence of this book.

Finally, the author would like to express sincere thanks for the interest to this topic and hopes that notwithstanding all the imperfections of the text, the reader will enjoy going through the book. All readers who notice errors or inaccuracies in the presentation of tasks, methods, and applications, and communicate these to the author[1] will receive a postcard of Varna in recognition of their contribution. The energetic critics of this work will also receive a small gift of appreciation from the author. Furthermore, given an explicit consent, their names will appear on the home page of this book[2] and in section Acknowledgements of its subsequent revised and extended editions.

Todor D. Ganchev

---

[1] Email address for providing feedback and error reporting to the author: tganchev@ieee.org
[2] Home page for the *Computational Bioacoustics* book, http://tu-varna.bg/aspl/CBbook/

Contents

Acronyms — XIII

1 Introduction — 1
1.1 Why should we care about biodiversity? — 1
1.2 Role of computational bioacoustics — 3
1.3 The benefits of computational bioacoustics — 5
1.4 Book organization — 7
1.5 How to use this book — 10
1.6 The home page for this book — 12

2 Why computational bioacoustics? — 13
Introduction — 13
2.1 Ancient roots — 15
2.2 Contemporary bioacoustics — 17
2.3 Computational bioacoustics — 22
2.4 Relations between bioacoustics and computational bioacoustics — 25
2.5 Research topics and application areas — 28

3 Definition of tasks — 32
Introduction — 32
3.1 One-species detection — 37
3.2 Species identification — 39
3.3 Multi-label species identification — 41
3.4 One-category recognition — 45
3.5 One-species recognition — 47
3.6 Multispecies diarization — 49
3.7 Localization and tracking of individuals — 52
3.8 Sound event-type recognition — 54
3.9 Abundance assessment — 57
3.10 Clustering of sound events — 58
3.11 Biodiversity assessment indices — 60
3.12 Practical difficulties related to real-world operation — 62

4 Acoustic libraries — 64
Introduction — 64
4.1 The training library — 68
4.1.1 TL organization — 68
4.1.2 Creation of training libraries — 71

4.1.3	Manual creation of TLs —— 72	
4.1.4	TLs based on clean recordings —— 75	
4.1.5	Challenges related to TLs' creation —— 76	
4.1.6	Computational bioacoustics in support of TL creation —— 77	
4.1.7	What if large TLs are not feasible? —— 77	
4.2	The acoustic background library —— 77	
4.2.1	What if large BL is not available? —— 79	
4.2.2	Computational bioacoustics in support of BL creation —— 80	
4.3	The validation library —— 82	
4.3.1	Computational bioacoustics in support of VL creation —— 85	
4.4	The evaluation library —— 86	
4.4.1	Computational bioacoustics in support of EL creation —— 87	
	Concluding remarks —— 88	
5	**One-dimensional audio parameterization —— 89**	
	Introduction —— 89	
5.1	Audio pre-processing —— 90	
5.1.1	Variable-length segmentation —— 92	
5.1.2	Uniform-length segmentation —— 93	
5.2	Audio parameterization —— 93	
5.3	Post-processing of audio features —— 96	
5.3.1	Statistical standardization —— 98	
5.3.2	Temporal derivatives —— 98	
5.3.3	Shifted delta coefficients —— 99	
	Concluding remarks —— 100	
6	**Two-dimensional audio parameterization —— 101**	
	Introduction —— 101	
6.1	The audio spectrogram —— 102	
6.2	The thresholded spectrogram —— 105	
6.3	Morphological filtering of the thresholded spectrogram —— 107	
6.4	MFCC estimation with robust frames selection —— 114	
6.5	Points of interest-based features —— 121	
6.6	Bag-of-instances audio descriptors —— 126	
	Concluding remarks —— 135	

7	**Audio recognizers — 137**
	Introduction — 137
7.1	Overview of classification methods — 137
7.2	One-species detectors — 143
7.3	Single-label classification with a single model per category — 148
7.4	Single-label classification with multi-classifier schemes — 151
7.5	MIML classification schemes — 154
	Concluding remarks — 159
8	**Application examples — 161**
	Introduction — 161
8.1	The ARBIMON project — 164
8.1.1	The ARBIMON Acoustics project — 164
8.1.2	The ARBIMON II project — 167
8.2	The AmiBio project — 170
8.3	The Pantanal Biomonitoring Project — 178
8.4	The SABIOD project — 181
8.5	The ENTOMATIC project — 182
8.6	The REMOSIS project — 184
	Concluding remarks — 185
9	**Useful resources — 189**
	Introduction — 189
9.1	Online audio data sets and resources — 190
9.1.1	The xeno-canto repository — 190
9.1.2	The Borror Laboratory of Bioacoustics Sound Archive — 191
9.1.3	The Macaulay Library — 191
9.1.4	The British Library Sound Archive — 191
9.1.5	Sound Library of the Wildlife Sound Recording Society — 192
9.1.6	The BioAcoustica repository — 192
9.1.7	The Animal Sound Archive at the Museum für Naturkunde in Berlin — 193
9.1.8	Australian National Wildlife Collection Sound Archive — 193
9.1.9	The Western Soundscape Archive — 193
9.1.10	The DORSA Archive — 194
9.1.11	The Nature Sounds Society — 194
9.1.12	The Acoustics Ecology Institute — 194

9.1.13	Sound resources not listed here —— **194**	
9.2	Software tools and services —— **195**	
9.3	Online information —— **198**	
9.4	Scientific forums —— **199**	
9.5	Technology evaluation campaigns —— **200**	
	Concluding remarks —— **203**	

Epilogue —— **205**
References —— **209**
Index —— **221**

Acronyms

DFT	Discrete Fourier transform
DRN	Dynamic range normalization
DWPT	Discrete wavelet packet transform
EER	Equal error rate
FA	False acceptance; false alarm
FB	Filter bank
FN	False negative
FR	False rejection
IBAC	International Bioacoustics Council
IoT	Internet of things
GFCC	Greenwood function cepstral coefficients
GIS	Geographical information system
GMM	Gaussian mixture model
GUI	Graphic user interface
HFCC	Human factor cepstral coefficients
HMM	Hidden Markov models
HTK	Cambridge HMM toolkit
LFCC	Linear-frequency cepstral coefficients
MFCC	Mel-frequency cepstral coefficients
SNR	Signal-to-noise ratio
STDFT	Short-time discrete Fourier transform
WSN	Wireless sensor network

1 Introduction

1.1 Why should we care about biodiversity?

Biodiversity is essential to the health of Earth's ecosystems. The consequences of biodiversity loss translate to impoverishment of environment, lower productivity, and lessening its capacity to support complex life, including the human species. The improved public awareness about these consequences made possible the prioritization of efforts aimed at the development of conservation planning policies. The last one brought forward the need of ecological impact assessment of all human activities, which indirectly contributes to the biodiversity preservation efforts (Rands et al. 2010; Collen et al. 2013).

Essentially, in the past decade, biodiversity preservation efforts gained some wider public and governmental support.[1,2] This is a good indicator of the awakened public consciousness about the seriousness of these problems, and the comprehension of associated risks and potentially devastating consequences on a global scale. If these risks are not managed appropriately and in a timely manner, further escalation of threat might pass the point of no return, which will trigger a chain of global catastrophic events that will probably affect the entire human civilization.

To this end, it has been undoubtedly proven that the efforts invested in biodiversity preservation, conservation planning, and environmental impact assessment bring positive effects to all human economic activities and contribute to improving the quality of life. In the same time, however, at present, the level of public support is limited to specific local problems and therefore is not always aimed at solving the global problems of biodiversity preservation.

This is because global biodiversity monitoring initiatives, such as the Global Biodiversity Information Facility (GBIF),[3] typically require monitoring of various families of animal species, inhabiting large territories, in order to assess

---

[1] The Intergovernmental Science-Policy Platform on Biodiversity and Ecosystem Services (IPBES), an intergovernmental body that implements assessment of the state of biodiversity in response to requests from decision makers, http://www.ipbes.net/

[2] The Strategic Plan for Biodiversity 2011–2020 provides an overarching framework on biodiversity for the entire United Nations system and all other partners engaged in biodiversity management and policy development. Cf. Aichi Biodiversity Targets, Strategic Goals A–E, https://www.cbd.int/sp/targets/

[3] The Global Biodiversity Information Facility (GBIF) is the biggest biodiversity database on the Internet. It is an international open-access data repository, which was established in the year 2001 with the aim to provide freely and universally accessible biodiversity information to

DOI 10.1515/9781614516316-001

the practical effects of certain measures and mitigation actions implemented in support of biodiversity conservation.

Therefore, the monitoring of certain key biodiversity indicators is a mandatory task of every effort in support of biodiversity preservation. Habitat protection and biodiversity preservation activities are profoundly dependent on the availability of rapid and precise biodiversity assessment survey methods and on the feasibility of continuous long-term monitoring of certain biodiversity indicators (Hill et al. 2005). These are mandatory components of every conservation action; however, neither of these are trivial tasks! Fulfilling these and other related tasks is challenging mostly because nowadays biodiversity assessment studies depend on the involvement of human experts on all stages of work. By that reason, biodiversity assessment and biodiversity monitoring actions are time demanding, expensive, and therefore not scalable.

The high cost of expert-based studies, and the enormous time effort required for the implementation of biodiversity monitoring and biodiversity assessment surveys, imposes limitations on their scope. By that reason, such studies are implemented only periodically, predominantly during daytime and favourable weather conditions, in locations that are less challenging logistically, and so on. These and other practical limitations currently narrow the spatial and temporal coverage of biodiversity-related studies and impede the efforts to understand the overall biodiversity dynamics on a global scale. All this significantly limits our chances to comprehend and model the underlying large-scale processes and to figure out and implement efficient conservation strategies.

Hence, it is highly desirable to provide strong technological support to biodiversity monitoring and assessment actions. Automation of certain tedious biodiversity monitoring procedures and ecological impact assessment efforts is seen as the only reasonable approach for addressing successfully the above-mentioned challenges imposed by long-term large-scale efforts aiming at biodiversity preservation.

Fortunately, technological tools and automation of data collection procedures started to gain trust in biodiversity assessment studies and were reported to enhance significantly the scope and temporal coverage of these surveys when compared to traditional human experts-based surveys. In that sense, it is already

---

everybody interested. GBIF is currently funded by over 50 governmental institutions and non-governmental organizations in over 35 countries around the globe. To this end, GBIF accumulates information for over 1.6 million species.

In order to facilitate potential data contributors and public data use, the GBIF website provides appropriate tools, support, and information about good practices in support of educators, researchers, institutions, and governments, and every interested citizen. Cf. The GBIF website, http://www.gbif.org/

acknowledged that a well-designed technological support could complement the work of human experts, and thus, facilitate the global efforts to preserve biodiversity and the vitality of life-supporting environment on the Earth.

1.2 Role of computational bioacoustics[4]

Sounds play an important role in nature. For instance, animals use sounds for communication in order to recognize and localize potential prey, to repel contenders, or to evade predators. Besides, sounds are emitted as a by-product of typical activities, such as locomotion, feeding, and so on. Therefore, animal sounds are considered an important information source in biodiversity assessment studies – sound propagates well in darkness, and when compared to eyesight is less affected by vegetation and obstructions. By that reason, most often animals are easier to hear than to see (e. g. due to their small size, camouflage, nocturnal activity patterns, etc.). These and other reasons established bioacoustic methods as major survey means and essential data contributors in biodiversity assessment studies (Pijanowski et al. 2011a, 2011b).

Terrestrial bioacoustic studies typically rely on passive monitoring methods, which remotely register the acoustic emissions of animals in their natural habitats.[5] The greatest advantage of passive bioacoustic methods is that they are less obtrusive when compared to traditional study methods, such as mark recapture, GPS tracking, mist netting, DNA analysis, and so on (Hill et al. 2005). Audio-visual survey methods (Jahn 2011) obviate the negative consequences of animal capture and any physical contact with animals, which helps to evade risks related to accidental death of captured animals. However, the audio-visual survey methods are human expert based, and therefore, their successful application depends

---

[4] The designation *Computational Bioacoustics* was first used in the book "Computational Bioacoustics for Assessing Biodiversity" (Frommolt, K.H, Bardeli, R., Clausen, M.), *Proceedings of the International Expert Meeting on IT-Based Detection of Bioacoustical Patterns*, December 7–10, 2007, International Academy for Nature Conservation (INA), Isle of Vilm, Germany (Internationale Naturschutzakademie Insel Vilm), BfN-Skripten, 2008, https://www.bfn.de/fileadmin/MDB/documents/service/skript234.pdf

[5] Terrestrial and underwater bioacoustics have equal importance to biodiversity preservation and share common history, signal processing, and machine learning methods. In the present book, we focus our discussion on sounds of terrestrial species that are in the human audible range. Many of the concepts and methods discussed throughout the book chapters are applicable to ultrasonic sounds, which play an important role in the monitoring and conservation of insects, bats, dolphins, and so on. The readers interested in underwater acoustic monitoring may refer to Frisk et al. (2003), Au and Hastings (2008), Popper and Hawkins (2012), Todd et al. (2015), and elsewhere with keywords *passive underwater (bio)acoustics*.

on the availability of highly skilled biologists and their extended presence on the field. The last does not facilitate long-term studies and scalability. Important advantages of passive bioacoustic methods are that
1. they are remote, that is, animals can be observed from a distance;
2. they do not cause significant disturbance to animal behaviours;
3. they do not depend on physical contact with the animals;
4. certain steps of data collection, management, and processing can be automated with the use of contemporary technology; and
5. observational data are recorded and can be stored for prolonged periods of time, analysed at a later stage, repeatedly used in long-term comparative studies.

To this end, bioacoustics already makes extensive use of technological tools. For instance, nowadays, autonomous recording devices are capable of collecting data over prolonged periods (weeks and months) without maintenance for battery or storage replacement. Furthermore, certain brands of equipment also possess the functionality to transmit wirelessly all, or some portion of, the recorded data in near real time or on demand. However, the data processing, analysis, and interpretation tasks largely depend on the availability (and the qualification) of experienced bioacousticians who need to inspect and/or listen to a certain amount of audio recordings.

The main point here is that although nowadays technology provides relatively cheap ways of collecting continuous recordings in various habitats, the workflow of data analysis and interpretation is high priced. This is because bioacousticians need to manually inspect and process all recordings (or in the best case in a computer-assisted manner) in order to carry out most of the tasks related to data analysis and interpretation. At the same time, a large-scale biodiversity monitoring project would require continuous recording over multiple locations, and therefore, the scale of effort and cost required becomes prohibitive – primarily due to the big data problem and the limited scalability of human expert-based study methods. The limited scalability is seen as the major obstacle that restricts the scope and time scale of biodiversity monitoring actions. In a typical present-day study, these are restricted to few particular species, monitored periodically in a relatively small area, over a number of short fieldwork missions.

In order to explain better the role of *computational bioacoustics*, in Chapter 2, we provide details on the historical roots of bioacoustics, briefly account to the transformation of needs, tasks, scope, and methods during its exciting evolution over the past 100 years, and then discuss its ongoing transformation to *computational bioacoustics*. In brief, the remarkable advance of information processing methods and communication technology in the past decade created the prerequisites for the emergence of a wide range of automated tools for data acquisition, transmission, storage, processing, and visualization. The conjunction of traditional bioacoustics with advanced information extraction and knowledge retrieval methods, communications technology, computer science resulted in the emergence of the new scientific discipline – *computational bioacoustics*.

Computational bioacoustics aims at scalable biodiversity monitoring. Therefore, it aims to develop automated methods and tools that provide the badly needed technological support so that problems related to big data acquisition, organization, processing, analysis, and interpretation are addressed in a scalable manner.

In the present book, we focus on topics of *computational bioacoustics* that are closely related to contemporary methods for automated acoustic recognition of species and sound events. These methods comprise the fundamentals of technological support to real-world applications, aiming at automated biodiversity monitoring and assessment, pest control, monitoring the spread of disease-transmitting insects, and other related applications where automated acoustic recognition of species is of primary concern. In Chapter 2, we briefly outline some applications where automation of sound recognition conveys the biggest impact to the current expert-based methods, and in Chapter 3, we define the main technological tasks that are in the focus of discussion throughout subsequent chapters.

1.3 The benefits of computational bioacoustics

Computational bioacoustics aims at scalable solutions, which facilitate the implementation of large-scale biodiversity monitoring studies. In the long term, these solutions contribute to the creation of advanced audio processing technology in support of successful coping with the global challenges of biodiversity preservation.

Computational bioacoustics aims to develop robust audio processing methods. Taking advantage of the fast progress of information technology, computational bioacoustics, among other tasks, aims to develop robust audio processing methods that successfully cope with the challenges of real-world operation and provide scalable tools for the benefit of biodiversity assessment and conservation actions. These methods and tools constitute the foundations of automated technological support that would permit scalability and would make feasible the implementation of unattended long-term biodiversity monitoring, acoustic pest control, audio information retrieval, and other relevant services for data analysis and interpretation. For instance, automated acoustic monitoring tools provide the means for capturing acoustic activity of species, which are rarely seen due to nocturnal activity or which occupy hardly accessible habitats, dangerous places,[6] and so on. Among the most important advantages of passive automated acoustic monitoring is that it is unobtrusive to animals and can be implemented continuously in 24/7 mode over extended periods of time. In this regard, *computational bioacoustics* well supports the concept of integrated archival of raw recordings, audio processing tools, the corresponding statistical models, and data analysis results. A great advantage of the *computational bioacoustics* methods comes from the fact that digital audio recordings collected by automated recording stations can be duly archived over extended periods of time, and when needed rechecked and reprocessed multiple times with different tools. In such a way, together with the original and processed data, one has the opportunity to store also the software tools with the specific settings used to obtain these results, together with the models created, and together with the outcome of the data analysis. Such a functionality facilitates reproducible research, offers opportunities for a significant improvement of the understanding about results obtained in previous studies, and assists for advances in methods and technology development. For instance, such a functionality facilitates studies aiming at a comparison between previous methods, tools, models, with results of new methods, tools, models, results, etc., and could offer valuable support to future studies.

All this facilitates the advancement of methods and technology development activities and provides convenient mechanisms to the research community to establish a long-term repository of shared resources and to foster collaborative environment. More importantly, however, is that all this permits biodiversity monitoring bodies and ecological impact assessment organizations to track previous data, history of analyses and conclusions drawn, efficacy of decisions and conservation measures implemented in support of biodiversity preservation goals. Such an organizational support is of a great value for the efficient management of resources and for planning of future activities. The enhanced transparency also strengthens the much needed public support, volunteering, and crowd sponsoring of biodiversity-oriented activities.

Computational bioacoustics contributes to automated retrieval and interpretation of information. The automated retrieval of information from soundscape[7] recordings or heterogeneous multimodal multi-sensor records, collected over

---

[6] Here, the notion of *hardly accessible habitats* or *dangerous places* refers to the perspective of an average human observer, and thus does not necessarily possess the attributes of an insurmountable obstacle.

[7] *Soundscape* refers to the combined acoustic perception of all sounds in a certain environment, including animal sound emissions, human activities, and environmental noise due to wind, rain, and so on.

extended periods of time, is highly desirable as it would allow to improve the coverage of biodiversity monitoring, assessment of environmental impact for certain human activities, pest populations control applications, and so on. The automated retrieval of information also facilitates the development of technological tools that would facilitate the interpretation of important acoustics events or the general trends of population dynamics and balance among species. The benefits of automated pest populations control also come mainly due to the possibility of continuous monitoring for evidence of infestations, the opportunity for implementing an adequate observation inside storage facilities, and improvement of chances to provide coverage of open fields at a reasonable cost.

Computational bioacoustics provides benefits to habitat management applications. In particular, the development of an advanced habitat management applications often requires functionality that supports prompt alerts about certain illegal and potentially dangerous events. Examples of such sound events are gunshots, forest fires, tree logging, unauthorized presence of humans, motorbikes, motor-powered boards, airplanes, and so on in protected areas. Many of these could be detected based on acoustic evidence. Evidence about unauthorized activities could provide important clues to a purposely designed expert system, and thus facilitate decision support in the long-term management of natural habitats and areas with some legal protection status.

In this book, we focus on methods applicable to acoustic monitoring of wildlife species, biodiversity assessment, and the acoustic detection of pests. We discuss the latest advances in computational bioacoustics that make possible some of the functionality desired in the above-mentioned applications to be incorporated in real-world technology. These advances make use of state-of-the-art signal and image processing methods and take advantage of well-proven statistical machine learning techniques. Some of these methods originally emerged for the needs of speech, audio, or image processing applications. Here, we discuss these methods from the perspective of computational bioacoustics and account for the specifics of automated acoustic recognition of species. Elaborations aimed at better coping with the challenges of real-world acoustic conditions and to requirements for continuous operation in 24/7 mode are discussed.

1.4 Book organization

This book is organized into nine chapters, which outline the foundations of computational bioacoustics, discuss recent advances, provide examples of successful applications of technology, and refer to further sources of information and useful resources. In short:

In Chapter 2, we start with a brief note on the historical roots of the cross-disciplinary research topic referred to as bioacoustics, and afterwards point out the challenges and the technology development milestones that made possible the emergence of computational bioacoustics. Next, we briefly outline the standing of computational bioacoustics as we understand it today, with respect to other scientific disciplines. In few short paragraphs, we discuss some inherent dependences of computational bioacoustics on infrastructure, hardware, and software resources, which are required for the practical deployment of real-world applications aimed at addressing local and global problems of biodiversity monitoring, environmental impact assessment of human activities, pest populations control, and so on. Although infrastructure and resources are often considered granted, and thus in many cases overlooked in discussions, in real-world applications many of these are not readily available, are limited in quantity and/or quality, or even could be out of question due to restricted resources. In that sense, Chapter 2 acquaints the reader with few basic concepts, which support the presentation of application-oriented projects (Chapter 8) that make use of *computational bioacoustics* methods.

In Chapter 3, we define the scope of computational bioacoustics and outline a variety of tasks it comprises. For each of these tasks, we discuss the problem from the end user's perspective and provide a brief description of the means and resources needed in order to address it. The last means that we put the emphasis on the type of input and output information interchanged between users and technology in each particular tasks. For each task type, we also discuss the prerequisite audio data sets used (and where applicable the type of expected tagging/annotation of data needed) for the model creation, for the fine-tuning of recognizers, and for performance evaluation. Considerations with respect to the practical use of these task-specific technological tools and services are also discussed.

In Chapter 4, we outline the purpose and significance of the various sets of audio recordings, referred to as *audio libraries*, which are required for the implementation of the research and technology development tasks discussed in Chapter 3. In addition, we comment on the main limitations and challenges related to the effort required for data collection and preparation, as this depends on the labour of highly qualified experts. The last and the amount of efforts needed for acquisition, preparation, and annotation of recordings make the endeavour prohibitive when efficient technological support and intelligent data annotation tools are not available. To a significant extent, the exposition in Chapters 3 and 4 provides the proper context and motivation for understanding the potential use of technology presented in Chapters 7 and 8.

In Chapter 5, we discuss the basics of traditional 1-D audio parameterization methods where the audio recording is first partitioned to a sequence of overlapping segments and then each segment is processed uniformly and independently of the others. In this category fall all methods that compute audio features directly from the time-domain signal and all methods that apply time-frequency transformation on a sequence of segments but then each segment is processed independently. The outcome of 1-D audio parameterization is a certain set of audio descriptors, which are computed for each audio segment without taking advantage of the information obtainable from the neighbouring segments. Next, the sequence of audio feature vectors is post-processed in order to incorporate prior knowledge or information about the temporal dynamics of descriptors, smooth the sequence of feature vectors via filtering out certain variability, and/or facilitate the statistical modelling via reshaping the distribution of values. In addition, the post-processing may help to compensate for the main disadvantage of the 1-D methods, which is that they compute static parameters that do not account for the temporal dynamics of audio descriptors over a chunk of audio.[8]

In Chapter 6, we overview some recent methods for 2-D audio parameterization that make use of image processing techniques applied on the audio spectrogram, which is considered as a greyscale image. These image processing techniques operate simultaneously on multiple columns of the spectrogram, aiming to eliminate noise, interferences that are well localized in frequency domain, identify regions of interest. In the implementation of these image processing methods, one can incorporate domain-specific knowledge directly in the audio parameterization process and thus optimize it according to the particular application needs. The last is in contradistinction from the 1-D methods, where such prior knowledge is typically implemented at the feature vectors' post-processing step. Another major difference is that the 1-D audio parameterization methods can be implemented to operate as online methods and are capable of working in real or near real time, that is, with a negligible delay with respect to the signal acquisition moment. The 2-D methods typically work offline or with a significant delay with respect to real time, mainly because they process multiple audio segments simultaneously and therefore require a longer buffering of the audio signal.

In Chapter 7, we overview present-day pattern recognition methods that are frequently employed in computational bioacoustic tasks. Technological solutions that incorporate either single-model or multi-model implementations of these tasks are considered. Single-label and multi-label set-ups for automated recognition of species and sound events are outlined. Specifically, we emphasize the main differences between single-label and multi-label approaches from technological perspective. Single-label approaches force hard decision about the

---

[8] In this book, chunk of audio, or an audio chunk, refers to a portion of the audio signal. A chunk is not bound to some predefined size, and duration may range from seconds to minutes.

(dominant) species that is presumed acoustically active in a certain chink of audio. The method may also estimate the timestamps of begin and end times for the dominant species. In contrast, the multi-label approach redefines the tasks of species recognitions so that all acoustically active species present in a recording are identified. The outcome of multi-label classification is a ranked list of species, which might also include some probability score for each species.

In Chapter 8, we overview the technological advances promoted by various research and technology development projects among which are ARBIMON,[9] AmiBio,[10] INAU Project 3.14, SABIOD,[11] ENTOMATIC[12]. The focus is predominantly on large-scale research and technology development activities, which demonstrate significant technological innovations and proof-of-concept demonstrators. The common context of these projects is that they handle large amounts of recordings collected in real-world conditions. These projects are interesting because they made the difference with respect to previous research, despite the hardware and software limitations at the time of their implementation. We briefly present the aim and scope of each project and discuss examples of tools and services addressing specific practical needs. These are considered the present-day success stories that demonstrate the applicability, the potential, and the limitations of computational bioacoustic methods.

In Chapter 9, we point out some essential references to other sources of information that the interested reader might wish to consider. These include links to (i) online repositories where audio recordings and software tools are made publically available; (ii) technology evaluation campaigns focused on scalable computational bioacoustic methods; and (iii) scientific journals, conferences, workshops, and so on. We also enumerate a short list of websites, which offer community support through discussion forums, questions and answers support, and some ready-to-use resources that can support future studies and technology development activities in the scope of computational bioacoustics.

Concluding remarks and outlook of trends for future technology development are provided in "Epilogue".

1.5 How to use this book

This volume is not intended to serve as a textbook for a particular academic course. As a multidisciplinary research area, computational bioacoustics covers

---

9 The ARBIMON project, http://www.sieve-analytics.com/
10 The AmiBio project, http://www.amibio-project.eu/
11 The SABIOD project, http://sabiod.univ-tln.fr/
12 The ENTOMATIC project, http://entomatic.upf.edu/

various topics of bioacoustics, biology, computer science, digital signal processing, ecology, information technology, machine learning. Terminology from these areas of science is used here without detailed introduction to the basics. In fact, we assume that the reader is already familiar with the foundations of digital signal processing and machine learning technology and has a basic understanding about the fundamentals of audio processing. Nevertheless, some chapters of this book could be considered a supporting material to students following undergraduate or postgraduate courses that cover topics related to
- Computational bioacoustics (Chapters 2–9)
- Machine learning (Chapters 4, 7, 8)
- Digital signal processing (Chapters 5 and 6)
- Image processing (Chapter 6)
- Application-dedicated courses concerned with audio processing methods and technology (Chapters 3–9)

In addition, PhD students researching in the area of computational bioacoustics, audio parameterization, information retrieval from audio, sound events recognition, or other topics related to audio processing technology might wish to consider reading Chapters 5 and 6 for deepening their knowledge and improving understanding on various facets of 1-D and 2-D audio parameterization. The formulation of tasks and the well-conceived application of state-of-the-art pattern recognition methods to computational bioacoustic tasks could be useful to students, who prepare a diploma thesis or a master's thesis on topics related to audio event recognition, acoustic recognition of species, information retrieval from audio recordings, and other related topics.

Computer science researchers, engineers, and software developers, who are not experts in computational bioacoustics, however, are interested in the development of applications, tools, and services, in support of periodic or continuous acoustic monitoring (24/7 mode) might want to go through Chapters 5–9. These chapters offer details on the basic methods and some less-known technological workarounds and trade-offs, and therefore may help in advancing their understanding. The same holds true for developers interested in applications that involve information retrieval or the automated detection of sound events.

We hope that the interested reader who goes through all chapters of this book would benefit not only on the side of methods and technology but also with an improved understanding of the role of computational bioacoustics in efforts for biodiversity preservation. Important highlights are biodiversity monitoring, biodiversity assessment, pest populations control, and applications related to monitoring for disease-transmitting mosquitoes. The improved understanding of tasks, methods, and the inherent limitations of contemporary technology would allow proper identification of opportunities for successful application. The discussion on tasks (Chapter 3), methods (Chapters 5–7), and the application examples (Chapter 8) facilitates the appreciation of the full potential of computational bioacoustics in providing support to future biodiversity preservation actions.

We hope that the present book, despite all imperfections and shortcomings, will contribute towards attracting and motivating more computer science researchers and young software- and hardware-oriented engineers to join the computational bioacoustics community. Such an increased involvement in the implementation of computational bioacoustics tasks will speed up the establishment of resources, tools, and services in support of information retrieval from soundscape recordings, improved data processing, and further developments in the area of sound event analysis. Such technological improvements will directly contribute to improved scalability of biodiversity studies and will facilitate future advances in biodiversity monitoring and assessment.

Each of the technology-oriented chapters in this book, for example, Chapters 5–8, points out essential references to relevant publications that provide further details on the implementation of computational bioacoustic methods and their application and validation on real-world problems. These references were not intended to provide a comprehensive list of all related work but instead we consider these a good starting point for further reading on the subject. The interested reader is encouraged to follow the numerous footnotes, which provide links to web resources and thus complement the main text.

1.6 The home page for this book

The interested reader may wish to follow the periodic updates on the home page[13] for this book. The home page provides links to further readings, related projects, a list of past technology evaluation events organized in support of the computational bioacoustics community, links to publicly available data sets, links and comments to useful software tools, and other resources covered in Chapter 9, or mentioned in the text. The author hopes that these links to publicly available resources will be useful to the wider readership, even if computational bioacoustics is not today's passion of this reader.

The author will be grateful for any feedback via email[14] or through the book's home page.[13] Any comments, suggestions, corrections, and bug reporting, or requests for further information and clarifications will be highly appreciated.

---

[13] Home page for the *Computational Bioacoustics* book, http://tu-varna.bg/aspl/CBbook/
[14] Email address for providing feedback and error reporting to the author: tganchev@ieee.org

2 Why computational bioacoustics?

Introduction

In order to explain better the scope and purpose of *computational bioacoustics*, we need to trace back its roots to *bioacoustics*, a cross-disciplinary research topic established a century ago. Because the scope and purpose of *early bioacoustics*, and thus the range of problems addressed, have changed significantly during the past century of development and advances, we will also need to outline the scope and the research problems studied by *contemporary bioacoustics*. Here, we shall make the important clarification that we perceive the transition to *contemporary bioacoustics* and *computational bioacoustics* in consequence of the natural enlargement of the research community, which is interested in animal sounds, biodiversity of species, and is concerned about the survival of Earth's ecosystems. Therefore, we need to recognize that in its first century of development bioacoustics gradually received increasing support from numerous scientific disciplines, partly due to the widening of its research scope and partly due to the involvement of researchers who work in other scientific topics and express interest to the problems addressed in bioacoustics. Such an increasing support is motivated by the potential of bioacoustic studies to contribute towards addressing problems with great social importance, problems that touch not only the quality of life of humankind but also correlate to the preconditions for the survival of human species.

In Section 2.1, we start with a brief note on the ancient roots of human interest to animal sounds. Next, we pay tribute to Dr Ivan Regen who is regarded as the founder of bioacoustics as a cross-disciplinary research topic that combines biology and acoustics, explain the evolution of scope and goals to *contemporary bioacoustics*, and clarify the needs that brought into existence *computational bioacoustics* (Section 2.2). Specifically, we first remark the main problems addressed by *contemporary bioacoustics* and summarize the present-day challenges and functionality demands imposed by its increasing role in biodiversity studies and its well-recognized support to biodiversity preservation. Among these is the fundamental requirement for scalability of biodiversity studies. Apparently, such a scalability is achievable only through the automation of data collection, processing, analysis, and interpretation. All this imposes the extensive use of automation tools and services provided by recent advances in signal processing and machine learning methods, computer science, communications, information technology, and so on, and therefore motivated the emergence of *computational bioacoustics*. However, the deployment of such an automated technology in real-world applications is conditioned also on achieving robustness of operation on

DOI 10.1515/9781614516316-002

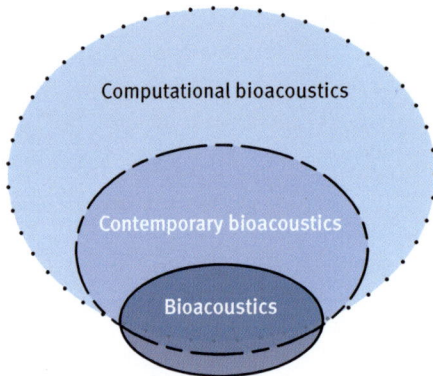

Figure 2.1: The broadening scope and shifting focus of early bioacoustics, contemporary bioacoustics, and computational bioacoustics – the introduction of technology facilitated the establishment of new methods, which contribute to addressing a wider range of problems related to biodiversity assessment and monitoring, and other problems with great social significance.

field recordings and coping with the challenges owed to the inherent variability of acoustic conditions in uncontrolled open-air environments.

The evolution of scope and purpose of *early bioacoustics*, *contemporary bioacoustics*, and *computational bioacoustics* can be illustrated with an enlarging oval (Figure 2.1). The ovals largely overlap; however, these are situated with different geometric centres due to the aforementioned shift of attention and change of goals. In other words, the focus of research in *early bioacoustics* was mostly on the sound production and perception mechanisms and certain relevant behaviours of species, while *contemporary bioacoustics* also gives emphasis to the ecology and biodiversity of species. The latter is of great importance to biodiversity assessment studies and efforts for halting the biodiversity loss.

As explained earlier, among the main goals of computational bioacoustics is achieving scalability of biodiversity monitoring and biodiversity assessment studies. For that purpose, computational bioacoustic methods rely on the extensive use of recent technological innovations, tools, and services that automate the tedious and expensive data collection, organization, and processing tasks. The author would like to acknowledge that the brief introductory statements here are oversimplified, and thus do not help in the comprehension of richness of interdisciplinary knowledge and methods used in *early bioacoustics*, *contemporary bioacoustics*, and *computational bioacoustics*. These will be discussed a bit further throughout Chapter 2, and then we will focus entirely on the tasks and methods of computational bioacoustics.

The main body of this chapter outlines the aims and scope of computational bioacoustics and the technology development milestones that made possible its emergence (Section 2.3). There, we briefly outline the standing of computational bioacoustics, as we understand it today, with respect to other scientific disciplines. Furthermore, in Section 2.4, we explain the relation between contemporary bioacoustics and computational bioacoustics and outline the key differences between these research areas. In fact, there is a significant overlap in their scope and methods, and both largely depend on the conjoint advances of methods and technology. In Section 2.5, we briefly mention real-world applications that motivated the emergence of computational bioacoustics and that heavily depend on its methods and technology. There, we briefly point out how computational bioacoustics could support each of these applications, given that all research challenges were addressed successfully. This brief account of required functionalities places the ground for the definition of the various computational bioacoustic tasks outlined in Chapter 3.

Throughout Chapter 2, we also emphasize some inherent dependences of computational bioacoustics on infrastructure, resources, hardware, and software tools and services, which are required for the practical deployment of real-world applications. While in research publications the availability of appropriate infrastructure and resources is often assumed granted, and thus frequently overlooked in exposition and discussion. In real-world applications these
1. might not be readily available,
2. are often limited in quantity and/or quality, or
3. are simply not feasible within the allocated resources in the project's budget.

In that sense, the discussion in Chapter 2 acquaints the reader with few essential concepts that facilitate the presentation of application-oriented technological solutions in support to recent research projects (cf. Chapter 8). These research projects address some central challenges of biodiversity monitoring, facilitate the environmental impact assessment of human activities, provide automated technology in support to pest control in agriculture, supply intelligent technological tools for monitoring the spread of disease transmitting mosquitoes.

2.1 Ancient roots

Since the early days of humankind, animals supply essential nutrition, clothing, and footwear, required for our survival. Alike most other predators, the early groups of *Homo sapiens* used to live near their main food source. That habitude used to guarantee emotional comfort and a convenient balance between hunting, leisure, and child rearing. Aiming to increase hunting success, early humans instinctively learned to scrutinize the behaviour of important prey species and to accumulate knowledge about their life cycle. Spiritual leaders and shamans mastered presentation techniques that helped them to mimic certain behaviours of prey and dangerous animals and possibly organized some ritual performances for the benefit of the tribe. In this regard, practical benefits are the training of young hunters, cultural and social commitment of youngsters, strengthening spiritual harmony among tribe members in provision of tribal survival.

Fortunately, these times are long gone and nowadays we have the option and possess the necessary knowledge and technology to manufacture food, clothing, and footwear from non-animal resources, which makes our survival less dependent on hunting. However, although the context is different, contemporary humans preserved the instinct for observing the behaviour of animals. Furthermore, people of most cultures make the conscious choice to share their modern homes with pets. Depending on the intent, people do it either out of pure curiosity, for entertainment, companionship, or professionally with a preset purpose, aiming to exploit a certain resource that animals provide, or for analysing potential threats, in support to efforts for preserving endangered species.

It is well known that the earliest evidence of human interest to the life cycle and behaviours of animals is documented in prehistoric cave paintings dated as early as approximately 40,000 BC. However, in the form of written records and books, the earliest known evidence of a systematic scientific study on the physiology, behaviour, and life cycle of animals is Aristotle's ten-volume[1] *History of Animals* (Aristotle 350 BC; Thompson 1910). There, Aristotle reports his observations and conclusions about the features and long-term behaviour for a wide range of terrestrial and marine creatures.

From the distance of time and the perspective of our modern science methods, the work of Aristotle is perceived as constrained to observations and certain ratiocinations, even though with the help of assistants he also reportedly tracked certain marked animals in order to investigate their behaviour over extended periods. The ten-volume series on *History of Animals* is an indisputable evidence of the human consciousness about the importance of systematic research, accumulation of data, logical thought, deep understanding and insights, and the needs for consolidation of the solid body of knowledge about animal biology and behaviours. These early books also illustrate the pronounced understanding of ancient civilizations about the essence of scientific observation approach, and most probably about its importance to survival of humanity.

Nowadays, 23 centuries after Aristotle, humankind is still driven by inherited instincts and curiosity, and therefore continues to invest time and efforts in activities aimed at the observing, collecting, taking measurements of animals, or merely registering their sounds, appearance, behaviours. Throughout the centuries, these activities were carried out with the tools available at the particular stage of development. For most of the time, these were only the human

---

[1] The authorship of *History of Animals* volume 10 is disputed as its content was argued to be controversial.

perceptual organs, imagination, memory, and certain imitation skills, in addition to few simple measurement tools, paper, and a pen.

The technological tools required for the accomplishment of sound recording, transmission, and the subsequent analysis of the time–frequency structure of animal sound events started to emerge only at the end of the nineteenth and the beginning of the twentieth century. Despite the limited functionality and low reliability of technological tools, these early developments opened opportunities for the emergence of bioacoustics.

Over the decades that followed, bioacoustics made use of gradually evolving tools and accordingly adapted its study methods. This broadened the scope of research problems that can be addressed successfully. In the second half of the twentieth century, bioacoustics overcame the initial predisposition towards descriptive studies, made a noteworthy progress towards biodiversity-oriented research, started addressing other complex problems of animal communication and behaviour, and shaped into what here we refer to as *contemporary bioacoustics*.

2.2 Contemporary bioacoustics

From prehistoric times, animal sound emissions have been used for the acoustic recognition of species and in certain cases for the recognition of particular animals.[2] However, only in the last two decades of the nineteenth century it became possible for wildlife sounds to be recorded with technological tools – shortly after the invention of the mechanical sound recording devices, these became a useful tool for the registration and reproduction of animal sounds. Specifically, during the last decade of the nineteenth and the first decade of the twentieth century, technological tools such as the mechanical phonograph[3] and graphophone[4]

---

[2] Biologists traditionally use acoustic clues in the identification of mammal, bird, anuran, or insect species. Hobbyist birdwatchers, ecologists, and so on, also develop excellent skills to recognize a great number of animal species solely based on their sounds. Reportedly, indigenous people, and people who live close to nature, are able to recognize particular individual animals based merely on their sound emissions.

[3] The invention of the phonograph dates back to the second half of the nineteenth century. The earliest recording preserved until today is dated 9 April 1860, and was made with a phonograph developed by Édouard-Léon Scott de Martinville in 1854. Later on, in 21 November 1877, Thomas Alva Edison announced his phonograph, which was able also to reproduce sound. (Source: *The New York Times*, Jody Rosen, 27 March 2008, http://www.nytimes.com/2008/03/27/arts/27soun.html)

[4] Volta and his associates introduced various practical improvements on Edison's phonograph, and by the year 1881, the recording was already made on wax cylinders. Volta and associates referred to their invention as graphophone. It was patented and commercialized after 1887. (Source: Wikipedia, https://en.wikipedia.org/wiki/Phonograph)

became commercially available and soon started to be used for the recording of animal sounds.[5] This allowed environmental and animal sounds to be reproduced elsewhere at convenient time for the purposes of entertainment, research, commercial incitement, and so on. Such a functionality opened the opportunity for sounds, which cannot be identified immediately on the spot where heard, to be reproduced and analysed later on.

In our time, Dr Ivan Regen[6] is widely regarded as the founder of *bioacoustics* – a cross-disciplinary scientific discipline that combines knowledge and methods from biology and acoustics. This recognition came for his methodical studies on insect sound communication in the early years of the twentieth century (Regen 1903, 1908, 1912, 1913, 1914). In brief, Dr Regen systematically carried out controlled experiments, making use of phonographic recordings and telephone devices in order to investigate the sound emissions, hearing, and behaviour of crickets and katydids, and scrutinized the experimental results by using statistical analysis methods.[7]

Thanks to the work of Dr Regen, early bioacoustics merged methods and knowledge from the areas of biology and acoustics, and furthermore, established new methods that were built on the use of recording devices, telephone communication, and few other technological tools available at that time. Some research methods were shaped in accordance with insights from ecology, and others were influenced by the development of telephony and other technological innovations in the area of audio recording, storage, reproducing, and so on (Figure 2.2).

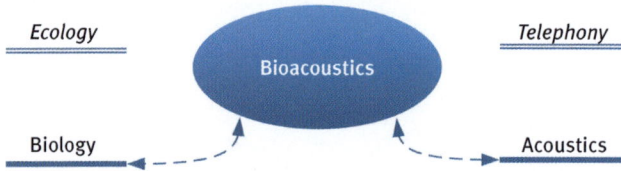

Figure 2.2: Early bioacoustics combined knowledge and methods from biology and acoustics.

---

5 According to Boswall and Couzens (1982), the earliest known phonographic recording of a bird song was made by Ludwig Karl Koch in the year 1889. In fact, Koch recorded a Common Shama bird in captivity as the early phonographs were not easy to operate on the field (Boswall and Couzens 1982). Reportedly, it took another 10 years or so until the first recordings of wildlife animals were made.
6 Dr Ivan Regen is also widely known as Prof Dr Johann Regen because for many years he taught in Technische Universität Wien, Austria, and published a significant number of his studies in the German language.
7 A detailed historical note of Dr Regen's work is available in Zarnik (1929).

Therefore, the introduction in practice of electrical and electromechanical equipment in support of the electrophysiological methods, used for the purpose of generation and measurement of sound stimuli, gave a strong impulse to the development of new methods in bioacoustics. A fascinating account of these advancements in methodology and enhancement of our understanding about animal sound communication is presented nicely in the overview on the history of animal bioacoustics (Popper and Dooling 2002). There, the interested reader will find out a brief history of bioacoustics, which covers a period of over 75 years, presented with an emphasis on the evolution of ideas and methods. The overview is based on analysis of publications in *The Journal of the Acoustical Society of America* for the period 1926–2002, which Popper and Dooling took as an indicative journal for the field of bioacoustics. They tracked the increasing interest to bioacoustics starting from mostly descriptive studies in the early years to the latest advanced comparative evaluations and multi-species surveys involving modern-day technological support and state-of-the-art methods. This overview article also credits some critical theoretical and experimental observations, which shaped the scientific area of bioacoustics to its present-day formulation.

Nowadays, contemporary bioacoustics studies[8] focus mostly on complex problems related to
1. sound production mechanisms,
2. hearing of animals,
3. sounds in taxonomy and systematics of species,
4. identification of species-specific sound events,
5. acoustic classification and identification of species,
6. analysis of multifarious sounds in the context of intra-species communication,
7. inter-species interactions between different families of animals (e. g. between birds and mammals), and so on.

In addition, advanced studies on the life cycle of certain species contribute towards behaviour analysis, ontogeny, learning, or investigate the impact of anthropogenic noise on these. Finally yet importantly, acoustic environment monitoring studies play an important role in support of focused efforts for decreasing the speed of biodiversity loss and assessment of ecological impact of human activities and/or natural calamity events.

This important role of contemporary bioacoustics, and the consequent rise of expectations to it, required extending the range of interdisciplinary studies.

---

8 The basics of animal bioacoustics are summarized in Rossing (2007), Chapter 17.

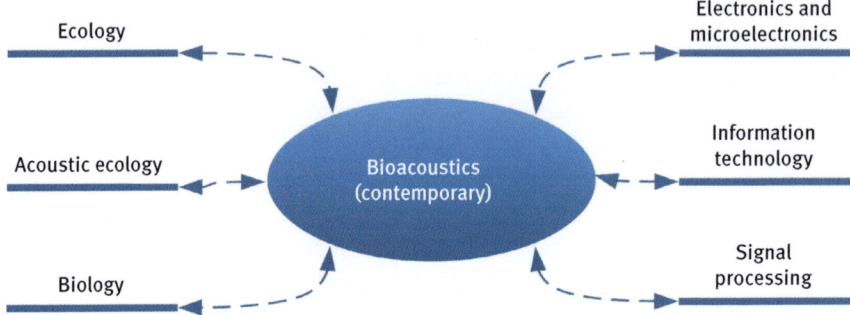

Figure 2.3: Contemporary bioacoustics: dependences on other scientific disciplines.

These included the elaboration of new research methods, which combine knowledge from ecology, acoustic ecology,[9,10] and biology, and depend on the advances of methods, tools, and services provided by certain technology development-oriented research areas, among which are electronics and microelectronics, signal processing, and information technology (Figure 2.3). Such a technological support extended significantly the scope and the range of problems that can be successfully dealt with the methods of contemporary bioacoustics, mostly due to the flexible manner in which data are collected, organized, stored, visualized, retrieved, shared, processed. This helped raise the role of bioacoustics and its broader use in biodiversity assessment studies.

From today's perspective, we could generalize that throughout the entire twentieth century the role of technology in bioacoustics was merely understood as providing support to particular activities, and therefore human expert involvement was the driving force behind every accomplishment. For instance, technology most often provided some functionality required for the establishment of specialized repositories of sound recordings and the corresponding metadata, as well as some tools in support of the visualization, manipulation, and annotation of sounds recordings and their spectrograms. In that sense, there were numerous studies which contributed towards

1. promoting new data collection methods, which make use of modern recording equipment, including recorders for the registration of mechanical

---

[9] Acoustic ecology is concerned with studies on the acoustic relationship between living beings and their environment, including the interference caused by human activities. A gentle introduction to acoustic ecology is available in Wrightson (2000).

[10] Acoustic ecology and bioacoustics are entangled with soundscape ecology. The merging of bioacoustics, acoustic ecology, landscape ecology, urban and environmental acoustics, behavioural ecology, and biosemiotics gave birth to soundscape ecology (Farina 2014).

vibrations, infrasound, sounds in the hearing range of humans, and ultrasounds; and
2. establishing new sound analysis methods, based on the temporal domain properties of audio, or on the audio spectrogram (i. e. the time–frequency domain distribution of energy of sound).

At the same time, studies that aimed directly at the development of sophisticated technological tools for automated acoustic recognition of sound emitting species, or at the automated recognition of certain species-specific sound events, were infrequent and of limited scope, in most cases covering just few species, or few genus at best. Moreover, for extended periods of time such studies were perceived loosely related to the objectives of bioacoustics and perhaps to some degree off subject.

However, a more important general conclusion is that until the beginning of the twenty-first century there was little progress towards the development and deployment of scalable automated methods for biodiversity assessment and monitoring. This is due to the lack of robust and scalable technological solutions, which are capable of coping with the challenges inherent in field studies. The implementation of a robust technology capable of operating in open-air, uncontrolled, acoustic conditions remains challenging. Yet, this is a basic functionality required for the implementation of unattended automated acoustic monitoring.

A major reason for this deficiency is the complexity of biodiversity assessment and monitoring studies, which significantly exceed that of weather monitoring, or other remote sensing networks. This is because the required technology for automated recognition of multiple species, which is essential in biodiversity assessment and monitoring, has to deliver accurate results and remain robust and reliable in varying acoustic conditions. The requirement for scalability to the needs of studies with different scope is even more challenging due to the amount of manual labour and expert resources required for the preparation of appropriate reference sound repositories and representative species-specific acoustic libraries, which are prerequisite for the development of automated acoustic recognizers.

Consequently, due to the limited technological support, the implementation of biodiversity surveys continued to depend entirely on the expertise of highly qualified biologists, that is, tasks were dependent on human labour and thus not scalable due to resource limitations. As we already emphasized, this dependence makes biodiversity surveys expensive, with limited duration and scope, and therefore not feasible for continuous monitoring of multiple habitats or for large-scale studies of key animal species over prolonged periods of time. In fact, these are not feasible merely from the perspective of contemporary bioacoustics; however, these appear possible from the perspective of computational bioacoustics.

2.3 Computational bioacoustics

Summarizing the discussion in Chapter 1, we highlight that *computational bioacoustics* emerged in response to the increased demand for scalable and robust automated technology, which is capable of coping with challenges inherent to real-world operational conditions, and which provides an essential support to biodiversity monitoring and biodiversity assessment studies. For that purpose, *computational bioacoustics* naturally stretched cross-disciplinary research activities and brought together present-day information and communication technology, signal processing, machine learning, with knowledge accumulated by contemporary bioacoustics and traditional natural sciences, such as biology and ecology (Figure 2.4).

It is comprehensible that *computational bioacoustics* became viable only after the recent emergence of high-performance technological tools provided by computer science and the advance of communication and information technology services. These together with the development of compact autonomous audio recording devices with large storage capacity and low power consumption, which became available only in the first decade of the twenty-first century, opened new opportunities for the automation of basic data acquisition and processing tasks. In addition, battery capacity improvement and the availability of cheap and efficient photovoltaic energy sources provided the means for prolonged autonomy

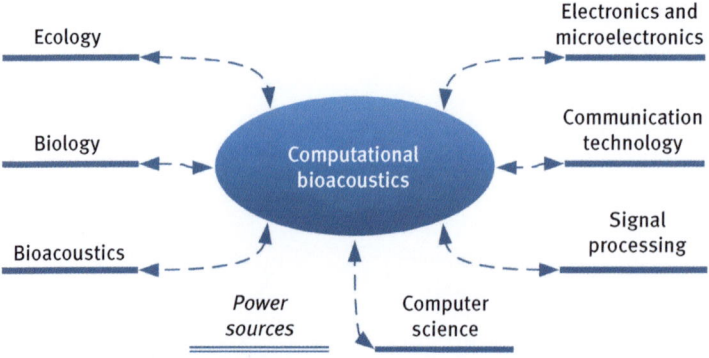

Figure 2.4: Computational bioacoustics: dependences on other scientific disciplines.

of automated recorders and reduced maintenance cost. These tools and services continue to evolve, taking advantage of the rapid progress in electronics and microelectronics, computer science, communication technology, Internet, which conditions the quick advances in computational bioacoustic methods and technology.

As illustrated in Figure 2.4, computational bioacoustic methods gain benefit from the advances in a broad range of technology-oriented and natural science disciplines. For instance, the increased efficiency of power sources (here we refer to the improved ratio of *battery capacity–weight* and the increased efficiency of solar panel power systems) and advances in electronics and microelectronics provided compact, robust, and user-friendly audio recorders, which are easy to deploy and maintain by staff members who are not technically inclined. Such recorders[11] typically offer autonomy of operation within a time span of several weeks to several months, entail low cost and low efforts for maintenance, possess options for telemetry and remote configuration of recording schedules and data acquisition settings.

Recent advances in electronics and microelectronics also placed the ground for enormous progress of communication technology and computer science, and made possible global wideband access to the Internet. At a local scale, off-the-shelf equipment allows deployment of custom-built wireless networks for data transmission. The advance of the Internet and the mobile phone industry accelerated the development of wideband wireless and cable communications and resulted in the establishment of global infrastructure. The handiness of such an infrastructure and the fast increasing throughput capacity of the Internet opened the opportunity of soundscape recordings collected at a certain microsite in the wild to be transferred in real time, or near real time, for processing and achieving purposes to the nearby data centre or to a distant data repository situated on another continent.

The data archiving functionality builds on methods and technology developed in computer science, and the current trend is shaped by the increasing availability of cloud-based storage and services. However, in computational bioacoustics, which often deals with tens or hundreds of terabytes of recordings, the use of cloud storage services largely depends on the quality and reliability of wideband Internet connections. This limitation is especially prominent in the wild areas with high biodiversity, where human economic activities have not stamped their footprint, or at least not to a great extent, and therefore wideband communication link might be absent. In contrast, in the areas with excellent

---

[11] For instance, the reader may refer to the Song Meter line of products of Wildlife Acoustics, http://www.wildlifeacoustics.com/products

communication infrastructure, and usually also with dense human presence and economic activities, biodiversity is poor. The former enforces computational bioacoustic projects to establish local storage centres with big data capacity or to seek other opportunities for data handling. The establishment of a large data repository might require a major increase of efforts, time, and resources when compared to the use of already existing cloud-based storage and services. Still it is important that contemporary electronics, microelectronics, and computer science also provide the hardware and software means for the establishment of such data archival and data processing centres at an accessible price –meaning that such centres are no longer the prerogative of few well-funded organizations here and there. Therefore, at present the insufficient communication bandwidth in areas with high biodiversity raises the need of such data archival and data processing centres. These centres are a prerequisite for the implementation of large-scale computational bioacoustic studies.

Furthermore, the availability of big data repositories, with hundreds of terabytes or petabytes of recordings within a single research project, requires certain efforts for data organization and structuring, as well as the availability of intelligent tools and services that provide access to contents. The development of such a functionality depends on the combined efforts of signal processing, machine learning, and computer science, which aim to provide the required intelligent tools and services for search in audio or multimedia recordings, content retrieval, and intelligent interfaces for visualization of dependences, evolution of parameters, trends, and prognoses based on the analysis and interpretation of data.

As nowadays computational bioacoustics is still in its early infancy, there is deficiency of ready-to-use tools and services in support of biology- and ecology-oriented large-scale studies. Furthermore, the rapid advancements in understanding, knowledge, and methods contributed by bioacoustics, biology, ecology, and other relevant disciplines largely influence the agenda of computational bioacoustic research. This is the major reason technology development to focus on specific applications, which are steered and prioritized by few large-scale research projects. In this regard, the demand for advanced functionality in certain computational bioacoustic tasks stimulates the development of methods in other scientific disciplines. Technology development activities, on their own, impose challenges related to certain technological and methodological problems, with fundamental or application-specific significance. All this contributes to broadening the scope of technology development activities and entails a wider collaboration between researchers from the computational bioacoustic community and other research areas.

In summary, the knowledge, infrastructure, tools, and services contributed by other scientific disciplines, as well as the internal research drive for advancement, foster the development of new computational bioacoustic methods. Taking advantage of the achievements and advances in a wide range of research disciplines opens new opportunities for successfully coping the challenges related to scalability and robustness of operation in field conditions. The availability of wideband communication infrastructure, powerful computational resources, and vast storage capacity facilities, which became handy and more accessible in terms of initial cost, and convenient services in terms of functionality and user-friendliness, fosters new opportunities for the advancement of computational bioacoustic methods. These methods aim at large-scale, continuous, automated, and simultaneous data acquisition at multiple locations, new data storage concepts, big data processing strategies, intelligent data visualization methods, and so on. In Section 2.5, we enumerate some biology- and ecology-oriented studies that would greatly benefit from the functionality provided by computational bioacoustic methods. In Chapters 5–7, we discuss the methods and technology that contribute towards successfully coping with the main challenges, linked to the automated processing of soundscape recordings. Examples of recent research projects that take advantage of these innovations are discussed in Chapter 8.

2.4 Relations between bioacoustics and computational bioacoustics

The aims and scope of bioacoustics and those of computational bioacoustics do not overlap entirely, so here we make an effort to explain more about the specifics. As said in Section 2.2, bioacoustics combines methods of biology and acoustics in order to study species, populations, and biodiversity. From a biology perspective, it can be perceived that biology seeks support in acoustics in order to establish methods that help us to gain certain information about the anatomy, physiology, taxonomy, behaviours, and life cycle of sound-emitting species. It also aims to assess various aspects of population size, density, dynamics, and trends by making use of human expertize and certain technological support. The focus here is on biology studies, and in this way, acoustics and technology have supporting roles. The researchers and practitioners working in the area of bioacoustics, also referred to as bioacousticians, predominantly have at least an undergraduate degree in biology and suitable training in some theoretical and practical topics of acoustics and informatics.

In the same time, it is much more difficult to see computational bioacoustics from a single perspective. It makes use of methods and technologies coming from electronics and microelectronics, computer science, signal processing, machine learning, communication technology, bioacoustics, biology, ecology, and so on, in order to implement large-scale biological studies.[12] Therefore, as a cross-disciplinary research area, computational bioacoustics can be seen from various perspectives. On the one side, we can say that engineers perceive computational bioacoustics as a specific application domain that makes use of advanced methods and technology. On the other side, ecologists, biologists, and bioacousticians perceive it as a biology-oriented discipline that receives certain technological support in order to implement biological studies. Both points of view are valid to a certain degree, yet both oversimplify and negligee a key point – at present, the successful handling of the computational bioacoustics tasks depends on research efforts in multiple scientific disciplines, often simultaneously! Therefore, computational bioacoustics is actually a cross-disciplinary research area, which addresses research questions related to scalability and robustness of methods and technology, which would allow real-world applications. This is only feasible through multidisciplinary research and the close teaming of researchers from various disciplines.

At present, it is unlikely that a single professional would be acquainted profoundly with the numerous research topics of computational bioacoustics needed for the implementation of a large-scale study. Moreover, even if so, a researcher might not possess all the expertise required, or at least not to the degree of proficiency that would allow her/him to

- design a statistical study plot for specific hypothesis test;
- select microsites for technology deployment;
- deploy an appropriate network of recording stations;
- organize the protocols and schedules for automated data collection;
- maintain the recorders over extended period of time;
- configure some software tools for data transfer over wireless infrastructure and/or over the Internet;
- manage and maintain a big data server or a cloud service;
- develop a database and appropriate data management tools with a convenient user interface;
- annotate recordings and prepare acoustic libraries,[13] needed for the development or automated acoustic recognizers for certain predefined species and/or sound events;

---

12 A brief note on these topics and the role of computational bioacoustics is provided in Section 2.5.
13 Details on the purpose of training, validation, and evaluation libraries are provided in Chapter 4.

- develop software tools for big data processing and automated content extraction, automated detection of species, and so on;
- evaluate the performance and deficiencies of the technological tools used, so that the study results obtained through such tools be given a proper interpretation; and
- carry out data analysis and evidence interpretation in order to draw conclusions about the biodiversity profile of a certain area, about the health of an ecosystem, about the conservation status of animal species, or make decisions about corrective measures in support of certain conservation action.

Depending on the study goals, various subsets of these activities might be implemented or extra tasks might be added.

Therefore, at present a major difference between contemporary bioacoustics and computational bioacoustics stems from the perception that a bioacoustician seems to possess the tools, the background, and the expertise needed in order to implement a study of certain complexity on her/his own. In the same time, the scale and scope of studies typically addressed in computational bioacoustics is prohibitively high and does not facilitate the human-expert based approach, but rather requires automation of certain tasks and teamwork. This is because nowadays it is uncommon that a single individual would possess profound knowledge in more than one of the aforementioned contributing disciplines (e.g. engineering, computer science, ecology, biology, etc.), and thus would have the proficiency to solve multidisciplinary research problems without a close collaboration with researchers form complementary fields of science.

For that reason, *one-man show* research missions are not common to scientific disciplines such as computational bioacoustics – a single-researcher undertakings rarely possess the manpower resources to provide sufficient depth and spatial coverage, unless these studies are focused on just one or a few particular species. Therefore, currently it is unlikely that a single person would be in a position to implement a large-scale study with respectable quality.[14] However, we do not exclude the possibility that in future when ready-to-use technology becomes available, there will be highly qualified professionals who are trained to implement large-scale studies, analyse, and interpret the outcomes on their

---

[14] This statement to some degree also refers to the limited perspective of discussion in the present book – the perspective presented throughout the chapters of this book is of an engineer (the author has a degree in electrical engineering). Thus, the author openly admits that he does not possess an appropriate background in biology, ecology, and so on, and also lacks the sensitivity and prudence to comprehend the essence of computational bioacoustics from the other valid perspectives: of a bioacoustician, a biologist, or an ecologist. The author hopes these limitations will not discourage the reader, and this book will be perceived useful, at least to technology developers and young researchers interested in methods of computational bioacoustics.

own. The last point is conditioned on the availability of sufficiently advanced off-the-shelf technology, so that one does not need to carry out hardware and software development tasks – for instance, when big data storage and cloud services for data processing become ubiquitous and more accessible. This is also conditioned on the introduction of university curricula on computational bioacoustics, along with a well-conceived practical training. Nowadays this is possible mostly through participation in the implementation of relevant research projects, summer schools, specialized training courses, and few other practical training opportunities. We speculate that in distant future computational bioacoustics will receive support from the advances in artificial intelligence if at that time biodiversity monitoring, assessment, and preservation are still of some concern.

In conclusion, we would like to emphasize that nowadays the success of computational bioacoustic studies heavily depends on the efforts of professionals with engineering, ecology, and biology background, and therefore the outcome is conditioned on a close collaboration and teamwork, with all the benefits and difficulties this might bring.[15]

2.5 Research topics and application areas

Computational bioacoustic research and technology development activities aim to automate certain time-consuming and tedious tasks, and hence free time and resources of highly qualified experts for the essential research, data analysis, and evidence interpretation responsibilities. Here, we refer primarily to repetitive manual labour and maintenance activities concerning the data acquisition, data

---

[15] Challenges in multidisciplinary collaborations, such those required in computational bioacoustics, could be trivial and, for instance, might be due to differences in terminology used, differences in the interpretation and application of scientific methods, in the priority of tasks, and so on. However, cultural differences, divergent agendas, dissimilar perception of research priorities, and commitment of researchers, when combined with various other aspects of human psychology and the potential mix between personal and professional relations, might raise serious issues during the implementation of projects. Thus, the quality of achievements in such cross-domain initiatives depends to a significant degree on the quality of teamwork and on the proper training and collaborative culture. The last is frequently underestimated, and if not addressed properly in the beginning of each project, especially when research will be implemented by a newly established consortium, it is afterwards learned the hard way. This might slow progress, lessen the involvement of researchers, and might require extra efforts for coping with the consequences "on the go" – alongside the project implementation. Thus, computational bioacoustics initiatives might greatly benefit from prior consensus on collaboration rules, especially if all collaborators agree to follow certain written rules, or established good practices and recommendations (Rosen 2007; Pritchard and Eliot 2012).

storage and organization, data management, indexing and searching in recordings, data annotation or tagging, manual recognition of certain species, acoustics events, and so on. Computational bioacoustic tasks, which can provide technological support to these activities, are outlined in Chapter 3, contemporary methods that facilitate the development of such technology are discussed in Chapters 5–7, real-world application examples are provided in Chapter 8, and in Chapter 9 we point out some relevant sources of further information.

In brief, considering studies that are concerned with sound-emitting species, the tools created by computational bioacoustics could potentially facilitate the automation of activities in various application areas, among which are the following:

1. *Biodiversity assessment and inventorying.* Traditional expert-based surveys allow a better scrutiny and a broader scope of methods and modalities, which contributes to a higher accuracy of results. On the other hand, technological support aims to make possible cost-effective implementations and continuous monitoring over extended periods of time, extend coverage to large territories and various habitat types, and so on. Human expert-based census studies rely on profound human expertise about a huge number of species and vastly benefit from human intelligence in order to pinpoint, identify, and understand the unexpected. However, in the long run, automated technology has the potential to combine models of numerous species and implement certain subsets of biodiversity assessment and inventorying tasks at a lower cost. The role of computational bioacoustics here is to create the technological means that would make possible the implementation of automated acoustic surveys and to develop tools and services that facilitate the work of human experts.

2. *Conservation status assessment of species.* An automated assessment of short-term and long-term population dynamics could facilitate the monitoring of specific species (e. g. with some particular conservation status), and thus contribute towards timely implementation of corrective actions when necessary. Computational bioacoustics could support these activities with purposely developed technology for automated collection of soundscape recordings, audio processing, and certain data mining and data analysis, based on acoustic evidence about the presence/absence of species and provide measurements about certain important aspects of acoustic environment in the corresponding habitats.

3. *Species richness and abundance estimation.* Automated estimation of the species richness in a certain area and the population dynamics of certain key species would allow a better understanding of the influence of natural and human-induced pressure. Such an improved understanding about the influence of these pressures would contribute towards achieving a more accurate

modelling of species population dynamics and potentially proper planning of focused conservation actions when required. Computational bioacoustics could well support these activities as it makes use of passive acoustic monitoring methods, which are unobtrusive to animals, do not distract or alter their activity patterns and behaviours, and do not cause harm to their habitats. Thanks to computational bioacoustic methods, evidence about richness and abundance is drawn solely based on the registered sound events, which are inherent to the activities of animals. Registration of sound activity is robust, entirely passive, and does not depend on the luminance level and on the presence of obstacles on the line of sight between the specific observation place and the position of the sound source.

4. *Monitoring for range shifts* of species. Human economic activities and industrial pollution entail shrinking of natural habitats, which decreases the opportunities for survival of wildlife species. An automated monitoring of the range shifts of migratory species, or changes in the activity patterns of nomad domestic species and the spread of invasive species, provides the means for disaster prevention and careful appointment of resources in support of focused corrective actions. Computational bioacoustic technology could support such monitoring efforts by a network of automated recorders, which register the acoustic activity patterns of particular species, send relevant information to data aggregation and processing facilities, and facilitate timely analysis of trends.

5. *Pest monitoring*. Early pest detection and pest density estimation are of significant importance for pest population control in human economic activities. For instance, continuous automated monitoring in grain storage facilities help for reduced use of chemicals, reduced cost, and for better preservation of grain quality. Furthermore, automated pest-monitoring applications could bring significant benefits in agriculture, for safeguarding crops, for timely protection of fruit trees, and so on. The benefits of passive acoustic monitoring by means of advanced computational bioacoustic methods are assessed in terms of reduced cost for chemicals and spraying activities, or in terms of higher profit, as organic agricultural production is in great demand, and therefore is priced higher on the market.

6. *Health-related applications*. Malaria, dengue fever, yellow fever, elephantiasis, and other infectious diseases from different microorganisms, including viruses (Zika virus, West Nile virus, etc.) and parasites, are known to be transmitted through mosquito bites. Other biting insects pose similar threats to humans and cattle as well. Computational bioacoustic methods and technology could provide the means for continuous automated acoustic monitoring of certain disease-transmitting species, continuous assessment of their spread

and density over certain territories, and thus contribute to timely planning of preventive actions against epidemic burst of infectious diseases.
7. *Automated information retrieval from soundscape recordings.* Large data repositories with millions of soundscape recordings and hundreds of terabytes or petabytes of audio have been accumulated over the years. The manual listening and analysis of these recordings is not feasible due to the enormous time and cost of such a task. Computational bioacoustics could provide automated audio content extraction tools to address these challenges. Such tools would facilitate long-term biological studies on biodiversity assessment, species behaviour profiles, and many of the above-mentioned tasks.

Certainly, our aim here is not to enumerate all research areas that can benefit from the advances in computational bioacoustics and list all potential applications that can take advantage of the automated tools it creates. Instead, we aim to categorize the tasks of computational bioacoustics (cf. Chapter 3) and illustrate the opportunities for providing solutions to practical problems (cf. Chapter 8) through automation and technological support (cf. Chapters 5–7) to some of these tasks as currently these are entirely dependent on the involvement of human experts. More specifically, we are concerned with those application areas, where convenient solutions are not currently available due to logistics limitations and due to various constraints related to the present-day dependence on the *in situ* presence of biologists and other qualified personnel.

3 Definition of tasks

Introduction

The scope of computational bioacoustics encompasses research and technology development activities focused on the creation of information retrieval methods, tools, and services in support of zoologists, ecologists, and managers concerned with the preservation of ecosystems. Depending on the specific research objectives, application requirements, and technology usage conditions, we discern various tasks of computational bioacoustics. In the following list, we briefly introduce some basic tasks. A comprehensive description of functionality and input/output information is provided throughout subsequent sections in this chapter.

- The *one-species detection task* is focused on the presence/absence detection, that is, it aims to verify whether acoustic emissions from a specific preselected species are present or absent in an audio recording. The outcome of such a process is always a binary decision ("Yes, the *species* was detected" or "No, *it* was not detected"), which indicates whether the sound emissions produced by the target species were detected or not in the given audio recording. Sometimes, the *one-species detection task* is also referred to as a *one-class classification* task to emphasize that there is only one species of interest. However, from the perspective of machine learning theory this task constitutes a typical two-class decision problem.
- The *species identification task* aims to discover if any species, from a set of preselected known species, is acoustically active in a given audio recording. This is a multi-class classification problem, which typically assumes that each recording contains sounds of a single species, or that the acoustic emission of the target species dominates over the background sounds.
- The *multi-label species identification task* aims to identify, enumerate, and list all species that are acoustically active in a given recording. Typically, the number of species, which are acoustically active in that specific recording, is not known in advance.
- The *one-category recognition task* involves the processing of soundscape recordings in order to select all portions where sounds from a specific broad category of species are present, for example, only birds, insects, anurans, bats, and so on. Such technology often serves as a gateway to reduce computational demand, or complexity, of the subsequent species (or sound event) recognizers.

DOI 10.1515/9781614516316-003

- The *one-species recognition task* aims to discover whether sound events originating from a certain predefined species are present in a soundscape recording and where exactly. For that purpose, the species-specific recognizer needs to discover the sound events originating from the species of interest and to determine the boundaries of each event, that is, to timestamp the start and the end of each event. The timestamps are usually formatted in terms of absolute time "HH:MM:SS.msec" in order to facilitate the hourly acoustic activity analysis of the target species.
- The *multispecies diarization task* aims to partition an audio recording into homogeneous segments, each containing sounds of a single species, to identify the species, to determine the timestamps of the start and end time of each sound event related to that species, and to create a diary of all sound events for a certain period of time. This allows the creation of temporal interaction diagrams, which facilitate the analysis and interpretation of acoustic activity patterns and interactions. The number of species in the recording might be known in advance (in the case of manual tagging before automated processing) but this is not a prerequisite, and in real-world set-up, it is usually unknown.
- The task acoustic *localization and tracking of individuals* aims to estimate the location and the movement trajectory for individual animals. This task assumes continuous or periodic acoustic activity of the tracked individual and requires recurring re-estimation of the spatio-temporal localization of individual animals for tracking their movement trajectories within an area. The area shape and size depend on the particular arrangement of microphones, their perceptive range, noise floor levels, and so on. This task is feasible for known and unknown species and individuals. However, the acoustic tracking of a particular specimen in the presence of akin sounds is more successful for individuals, which are known in advance and are explicitly modelled. Such locality and tracking information contributes to the analysis of species-specific activity patterns, behaviour recognition, abundance estimation, and might be used in population density assessment studies.
- The *sound event-type recognition task* aims to discover whether sound events from a predefined species-specific repertoire are present in a given audio recording and where exactly. This includes all types of sound emissions, such as calls, call series, songs, and so on, included in the typical repertoire of individuals of a certain species. When the information about these sounds is associated with a specific context, the sound event-type recognition task contributes information to the abundance assessment, activity pattern recognition, and behaviour recognition tasks.

- The task *clustering of sound events* aims to group together sound events based on predefined similarity criterion or criteria. This functionality could be of great help for the automation of species annotation and sound event-type annotation, which currently is performed manually by human experts. The task *clustering of sound events* also allows the implementation of *query-by-example* and *query-by-semantics* functionalities, so that one can search for a specific sound entity in a recording, in a selected subset of recordings, or in an audio repository. Entities could be of different nature (a call, call sequence, song, a subunit of a sound event, etc.) and might, or might not, be necessarily associated with a meaningful name. However, in any case these sounds should bear certain similarity features, based on which these will be selected. For particular types of sound events with less-complex time–frequency structure, it is possible to implement the query-by-example functionality, that is, query in the style *"search for this chunk of sound in this soundscape recording, or in a subset of audio recordings that meet the predefined criteria"*.
- The task *species abundance assessment* in a certain area aims to estimate the number of individuals of the preselected target species. Such an estimation is based on the recognition of sound events produced by single individuals, duets, and chorus. This task provides raw data for population density estimation and other related tasks.

While some of these tasks directly aim to develop methods, tools, and services for computer-supported processing and analysis of data, others are important components of future fully automated systems. Both will possibly find their use in applications related to (i) acoustic pest detection, (ii) ecosystem health assessment through acoustic activity monitoring, (iii) acoustic biodiversity inventorying studies, and so on. Categorization of the above-mentioned tasks with respect to the number of classes involved in the data modelling and decision-making process is shown in Figure 3.1.

The *one-species detection*, the *one-species recognition*, and the *one-category recognition* tasks are typical two-class decision problems. The *species identification*, the *multi-label species identification*, and the *sound event-type recognition* tasks are typical multi-class decision problems, concerned with retrieving information about which species are vocally active in a given recording or what kind of sound event was emitted. Such information could be useful in automated census studies. In addition, such information can also be related to any evidence derived through the various audio indices for rapid assessment of biodiversity (Sueur et al. 2008, 2014), which reveal the overall structure of a soundscape at a certain moment of time.

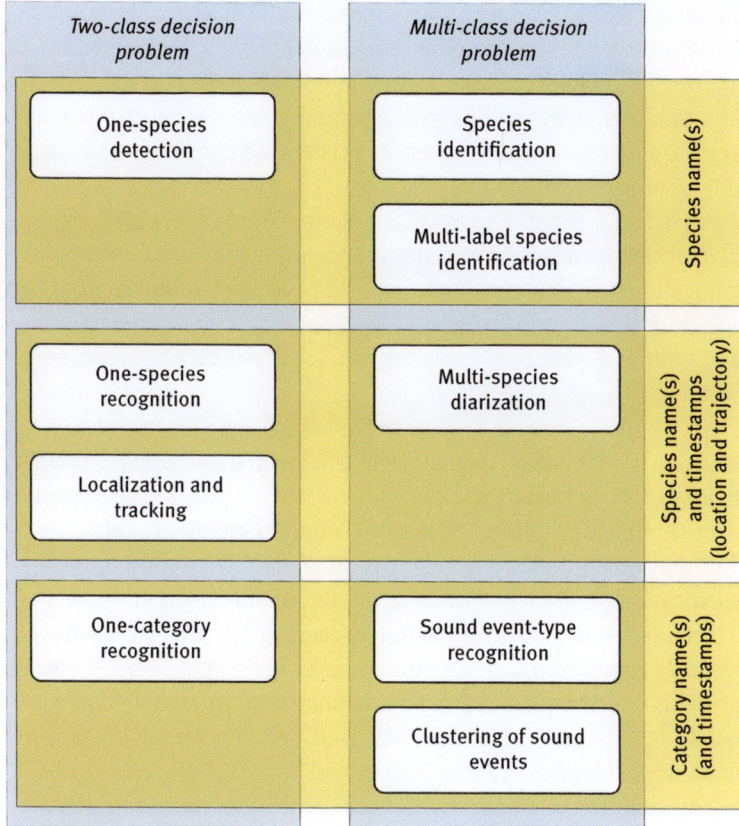

Figure 3.1: Categorization of computational bioacoustic tasks depending on the number of classes involved.

With respect to the type of information on which the above-mentioned tasks focus, we can distinguish the following three broad categories:
1. *Tasks focused on the identification of acoustically active species in a certain recording.* In this group fall the *one-species detection*, the *species identification*, and the *multi-label species identification* tasks, which aim merely to investigate the presence/absence of one or several known species for which there are either etalon example(s) or purposely developed model(s). These models are prebuilt in advance for known species, according to the objectives of the specific study. In the *one-species detection* and the *species identification* tasks, for each recording we designate only the species with predominant acoustic activity. In the *multi-label species identification*, we enumerate all species that are acoustically active in the given recording.

2. *Tasks aiming at the identification of sound events and their temporal boundaries.* These sound events are considered to originate from a predefined list of known species for which there are prebuilt models. In this category fall the *one-species recognition*, the *multispecies diarization*, and the *localization and tracking* tasks, which besides the species name also aim to estimate the start and end times of each audio event of interest.
3. *Tasks concerned with the detection of sound events grouped with respect to certain similarity criteria and the temporal boundaries of each event.* In this category fall the *one-category recognition*, the *sound event-type recognition*, and the *clustering of sound events* tasks. Besides the class name determination, these tasks also aim to estimate the start and end times of each audio event of interest.

The main difference between these categories is that the first only aims to confirm the presence/absence of the target species, while the second and the third are also concerned with detecting the boundaries of the corresponding acoustic events and the timestamps of their start and end. In addition to the species name and the timestamps, the *localization and tracking* task also aims to estimate the spatial location[1] of an individual. The movement tracker is typically combined with a one-species recognizer, which is particularly useful when there are other loud sound sources in the receptive field of the microphone array. Simultaneous tracking of multiple individuals is a quite complex task, which is usually reduced to tracking the individual trajectories independently, or in some advanced methods with the exchange of information among the individual trackers.

Furthermore, the third category of tasks is not concerned with the recognition of species but aims at grouping sound events with respect to certain similarity criteria. For instance, in the case of the *one-category recognition* task sound events from a specific category (insect sounds, anuran sounds, bird sounds, bat sounds, etc.) are selected. In the *sound event-type recognition* task, the focus is on discerning among the various calls, call series, or songs, which are emitted by individuals of a certain species. This information is typically linked together with contextual information, which could serve for analysing activity patterns and behaviours of single individuals or populations of species. In this category, we also include the various supervised or unsupervised sound event clustering tasks, which allow grouping of sound events with respect to certain similarity criteria, or tasks, which search for sound events by example, by semantics, or by some characteristic trait.

---

[1] In fact, most often the localization task is reduced to estimating the direction of arrival of sound relative to the position and the orientation of the microphone array.

It is worth mentioning that all the above-mentioned tasks differ mainly in the manner the species-specific or category-specific models are created and used, which depends on the requirements of the particular application. However, in general, there is not much difference in the underlying audio processing and pattern recognition methods involved in these tasks.

Finally, all species recognition tasks, which are defined as a multiple-class decision problem, can be reduced to a number of two-class decision problems, which brings some practical advantages during model retraining and extending the number of modelled categories.

3.1 One-species detection

The *one-species detection task* consists of inspecting whether acoustic emissions from a specific predefined species are present in an audio recording or in its preselected portion. The *one-species detection* task is also referred to as a one-class classification task, which emphasizes that there is only one species of interest. The outcome of the *one-species detection* process is a binary decision: "Yes, sound emissions produced by the target species are present in the audio file" or "No, the target species was not detected in the audio file." Such information could be useful, for instance, when one needs to retrieve all files recorded within a given time period, which contain acoustic activity of a certain species.

In general, it is feasible to make use of a template-matching approach that relies on comparing the input with an etalon. Depending on the application scenario, the etalon could be an instance of a specific call type, call series, song, part of a song, and so on. A final decision is made based on an empirical threshold applied on the score computed after aligning the etalon and the input recording.

Alternatively, one can build a single statistical model that represents the acoustic characteristics of the target species. Next, each input is matched to the model and a decision is made after applying a threshold on the score computed through some similarity measure, likelihood, or probability computed through that model.

However, both the template-matching approach and the single-model statistical approach do not perform well on noisy recordings. This is because these approaches do not incorporate a reliable mechanism for modelling the time-varying operational conditions. Therefore, in practical applications the *one-species detection* is implemented as a two-class decision problem where we build two models – one for the target species and the other for everything else, that is, for the rest of the world. In the ideal case, the latter is an exhaustive model, which models all possible backgrounds and interferences, and thus provides a mechanism for

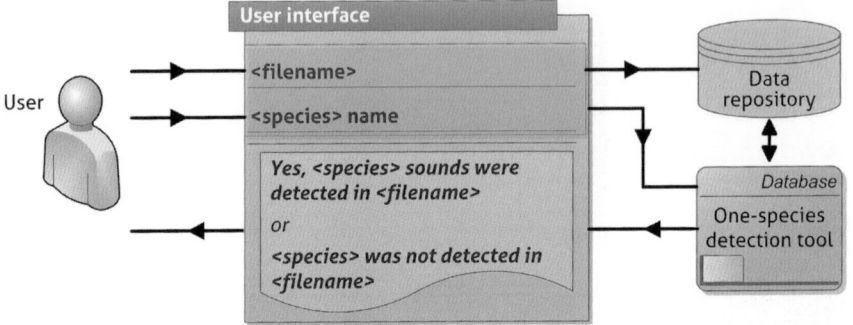

Figure 3.2: Conceptual diagram of the one-species detection task.

compensating the variability due to dynamic environmental factors and helps in improving the overall robustness of the one-species detector tool.

The conceptual diagram of the *one-species detection* task (Figure 3.2) specifies the input and output data types, common to all species-specific detectors regardless of their implementation. In fact, the user needs to provide the name of the species of interest, referred to as *target species,* and the name of the audio recording to be tested. The species name is linked to the name of a prebuilt acoustic model through an indexed database. When the input <filename> contains wildcard symbols, and by that reason there is more than one recording in the database repository that matches <filename> for the specified time range or interest, each audio file is evaluated separately and a binary decision is made on per file basis.

In statistical approaches for *one-species detection*, the hypothesis represented by the target species model competes against the alternative hypothesis, where the alternative hypothesis represents the acoustic environment typical for the particular recording site. The acoustic models representing these two hypotheses are built from purposely selected audio recordings. The former model is species-specific and therefore is created based on a species-specific training library (TL) and the latter one on an acoustic background library (BL). The TL is composed of tagged audio segments containing sound events of the target species. The BL contains recording from the particular recording site, which is manually screened to exclude audio segments containing the target species. A validation library (VL) containing tagged audio segments, among which annotated sound events of the target species, is used for fine-tuning the adjustable parameters of the one-species detector. Performance evaluation is carried out with the help of an evaluation library (EL) that consists of tagged recordings, which may or may not contain sound events of the target species. The specific content and build of these acoustic libraries are discussed in Chapter 4.

A comprehensive description of one-species detector based on the state-of-the-art statistical method is presented in Section 7.3. Related work on one-species detection is available in Potamitis et al. (2009), Bardeli et al. (2010), Graciarena et al. (2010), Ganchev et al. (2015), and Sebastian-Gonzalez et al. (2015).

3.2 Species identification

The *acoustic species identification* task aims to discover whether any species from a set of predefined known species is acoustically active in a given audio recording. This is a multi-class classification problem where we typically assume that each recording contains sounds of a single species or that the acoustic emission of one species dominates over all other sounds.

Obviously, the *species identification* task requires the creation of a species-specific acoustic model for each species of interest. During operation, the species identification tool scores each input recording against all existing models and then a final decision is made. Depending on the pattern recognition technique used, the final decision is made based on the estimated degree of similarity between input and models, likelihood scores, or probabilities, which are computed for each species-specific model.

The *acoustic species identification* could be defined in an open-set or in closed-set set-up. A closed-set species identification can be performed only in controlled set-up, for example, by making use of sound recordings collected in laboratory conditions, where all sounds come from a known species for which there are prebuilt models. In that case, the automated identification process is forced to tag the input recording by selecting one species among all that were modelled. In the closed-set species identification, it is typical that the number of models is equal to the number of species we aim to identify. It is also possible to have more than one model per species. However, when the number of species is high, the latter might be prohibitive due to increased computational demands.

When the application set-up imposes that the audio recordings to be processed may possibly contain dominant sounds from species, which are not intended to be modelled explicitly with a species-specific model, we refer to it as open-set species identification. In that case, the outcome of the automated identification process is either the species name, selected among all modelled species, or a notification message in the sense "None of the known species was detected in recording <filename>" (Figure 3.3). When compared to the closed-set species identification, the open-set set-up requires extra efforts (and often a properly adjusted threshold) in order to decide whether the input belongs to any of the known species for which there are prebuilt models, or not.

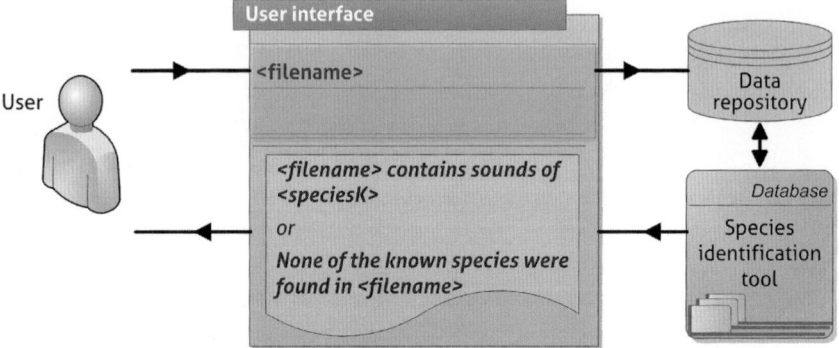

Figure 3.3: Conceptual diagram of the acoustic species identification task.

In the open-set species identification set-up, it is common to reduce the multi-class decision problem to a number of two-class decision problems. Each two-class decision component acts as an expert, which is tuned to detect one particular species. This requires that the outputs of all species-specific experts be post-processed for resolving conflicts before making a final decision. The main advantage of transforming the multi-class decision problem into a number of two-class decision problems is that when we need to retrain a model of a certain species, or to add a new model to the set, we do not need to rebuild all other models. Disadvantage is the increased computational demand during operation due to some redundancy and extra processing steps. However, the independence of species-specific models in the two-class scheme facilitates distributed computing, and thus the overall computation time for the multiple two-class problems might be smaller when compared to the direct implementation of the multi-class scheme.

Collectively, the open-set and the closed-set species identification tasks are also referred to as classification of audio recordings. This might be useful in tools for automated sorting of audio recordings. Related work is reported in Ganchev and Potamitis (2007), Bardeli (2008), Trifa et al. (2008a), Henríquez et al. (2014), and so on.

In the species identification task, the user only needs to provide the name of the input recording (Figure 3.3). After processing it, the species identification tool returns the name of the acoustically active species, choosing one among all known species for which there are prebuilt models. When the input recording does not contain dominant sound events from any of the species represented with species-specific model, the tool will either categorize the sound event in one of the known species (in closed-set set-up) or will tag it as "species unknown" (in open-set set-up). The latter would typically mean that the user receives notification in the sense of "None of the known species were found in <filename>."

In the *species identification* task, the creation of acoustic models depends on the availability of species-specific TL for each class. The TL would ideally contain recordings with loud acoustic emission of a single species per file or at least the species of interest needs to be dominant with sound amplitude high above the noise floor. A discussion of the practical difficulties associated with the creation of high-quality TL is available in Chapter 4. When the amount of audio recordings in the TL is insufficient and does not facilitate the creation of robust species-specific models, the identification accuracy of the entire system is degraded. One way to cope with this problem is first to create a general acoustic model and then derive all species-specific models through adaptation with species-specific training data. The general acoustic model may represent the operational conditions of the environment, in which case it is built from an acoustic BL (cf. Section 7.3). For that purpose, the BL has to contain a large amount of audio recordings captured in the operational environment. The BL has to be cleaned appropriately so that is does not contain acoustic events of any species that has to be identified.

Alternatively, the general acoustic model may represent a neutral entity that is positioned (in the acoustic space) equally away from all species of interest. Under certain conditions, such neutral entity can be built from a weighted mix of descriptors extracted from recordings of multiple species. However, the use of such a neutral entity is controversial and is rarely preferred to acoustic background model. Still it provides a satisfactory solution for the cases where acoustic BL is not available or if its development is not feasible with the available resources.

Once the species-specific models are trained, the open-set identification may require VL for tuning the adjustable parameters of the system, for example, decision threshold(s) for categorizing the input audio to "species unknown". Finally, the identification accuracy is evaluated based on an EL. Both the VL and the EL need to be representative of all species of interests and of the environmental conditions in which the species identification system will operate.

Due to practical difficulties related to the creation of TL, BL, VL, and EL, most studies on *species identification* are restricted to clean recordings collected in laboratory conditions (Chesmore 2001; Chesmore and Nellenbach 2001; Ganchev and Potamitis 2007) or to the identification of only few species when recordings are collected in field environment (Chesmore 2004; Trifa et al. 2008a; Huang et al. 2009; Bardeli et al. 2010; Henríquez et al. 2014).

3.3 Multi-label species identification

The *multi-label species identification* task aims to identify all acoustically active species present in a given audio recording or a chunk of it. This task does not require that all audio events are timestamped and labelled individually to the species they belong but instead the automated classifier simply outputs a list of all species which were discerned in that audio chunk.

The main difference with the *species identification* task (Section 3.2) is that here we no longer impose the assumption of one dominant species per audio segment. Therefore, the multi-label species identification requires that we separate, recognize, and enumerate all species in an audio segment without even knowing their total number. The last renders *multi-label species identification* more appropriate to the needs of automated acoustic biodiversity monitoring when compared to the tasks discussed in Sections 3.1 and 3.2.

Obviously, the *multi-label species identification* is quite a challenging task as acoustic events of different species often coincide in time and overlap in frequency range. Furthermore, due to different distances to the microphone and/or differences in the sound production mechanism, body size, and other species-specific features, sound events have quite a different range of amplitudes. The last imposes iterative processing of the audio stream, each time eliminating the loudest sound event and then seeking for the next one and/or the use of purposely developed method for 2-D separation of sound events in the time–frequency space (cf. Chapter 6) during the audio parameterization stage.

A direct approach for *multi-label species identification* based on the multi-instance multi-label classification paradigm is discussed in Section 7.5. This classification method typically operates with sound event features extracted in the time–frequency space, that is, combined with 2-D segmentation and feature extraction methods (Chapter 6). It has been demonstrated that the multi-instance multi-label classification outperforms all other methods in several technology evaluation campaigns (Lasseck 2013, 2015; July et al. 2015; CLEF- 2016).

By that reason, both the direct and iterative approaches for multi-label species identification operate offline, with prerecorded audio recordings or with a buffer of few seconds of audio. Therefore, these introduce delay of at least several seconds. However, such a delay and even offline operation are perfectly fine for applications such as biodiversity assessment, acoustic inventorying of species, and so on, where real-time operation is not required.

Similar to the species identification task, the *multi-label species identification* is a multi-class pattern recognition problem which can be reduced to a number of single-label decision problems. The last is conditioned on the feasibility of a reliable segmentation of all individual sound events existing in an audio chunk. Specifically, we aim to obtain a set of sound events, each corresponding to some class label, which are separated in time–frequency domain.

A major difference in the *multi-label species identification* when compared with the *one-species detection* (Section 3.1) and the *species identification*

(Section 3.2) tasks is that here the TL may no longer consist of recordings that contain a single dominant species per file. For instance, a recent technology evaluation campaign (e. g. The LifeCLEF2015 Bird Identification Task,[2] Multi-label Bird Species Classification competition NIPS4B[3]) provided a *train data set* that contains recordings with more than one species per file. There was no annotation of the temporal boundaries of each bird vocalization but only the label of the dominant target species and a list of other species heard in the background. This imposes another level of difficulty during training because the machine learning methods have to automatically segment, detect, and combine pieces of incomplete information that is not structured well and that might be scattered.

One great advantage of the *multi-label species identification* task is that it relaxes the requirements to the acoustic TL when compared to the tasks discussed in Sections 3.1 and 3.2. First, here the individual recordings in the TL may contain sound emissions of more than one species per file (most soundscapes and field recordings are like that). Furthermore, there is no need of manual data annotation for specifying the temporal boundaries of sound emissions of the target species or the requirement for selective manual editing of interferences from other sources, competing with the target species.

Indeed, the *multi-label species identification* task benefits from any information about the number of species that are acoustically active in each file and from having information about the species names. However, this is not a mandatory requirement, and therefore the TL may not contain this information either. Furthermore, it is not required that the exact order of species vocalization in the file or the temporal boundaries of their sound event are known. Thus, the preparation of TLs may become quite simple – it will be a collection of certain audio recordings of the target species, and in some cases, it may incorporate information about other species that are acoustically active in each file. These relaxed requirements mean a great reduction of the time and effort needed for the preparation of TLs, known as one of the major bottlenecks in the technology development process. We deem that once the *multi-label species identification* technology becomes mature and reliable, it will facilitate the implementation of large-scale species identification studies. Such studies are presently impeded by the prohibitive amount of resources needed for the preparation of species-specific TLs for many species.

Although timestamps are not necessarily required and the output could be simply the list of species present in an audio recording, the *multi-label species*

---

[2] LifeCLEF2015, http://www.imageclef.org/lifeclef/2015/bird
[3] NIPS4B, https://www.kaggle.com/c/multilabel-bird-species-classification-nips2013

identification task often needs to make use of methods for the automated detection of the temporal boundaries of each acoustic event in the train library and to enumerate all species present in each recording. This is what usually needs to be done during processing of soundscape recordings, which makes this task more appropriate to field research studies when compared to the *one-species detection* (Section 3.1) and the *species identification* (Section 3.2) tasks.

Similar to the *species identification* task, here the user only needs to provide the file name of the input recording (Figure 3.4). Next, the multi-label species identification tool discovers whether any of the known species, for which there exist prebuilt models, are acoustically active in the given audio file. The outcome of this process is a list of species which were detected to be acoustically active in the specific recording. If the audio file does not contain sounds from the species for which acoustic models are present, then a notification message will inform in the meaning of "*None of the known species were detected in recording* <filename>." Whenever the user input <filename> contains wildcards, and by that reason, there is more than one match in the data repository, each file is processed separately.

Recent developments in different aspects of the *multi-label species identification* task were reported in Briggs et al. (2012, 2013a, 2013b), Lasseck (2013), Goeau et al. (2014), Glotin et al. (2013c), Leng and Tran (2014), Menćia et al. (2013), Potamitis (2014), and Lasseck (2015). These studies built on advances in 2-D audio parameterization and machine learning methods, which were developed in the last decade for the needs of image processing and were afterwards successfully adapted for the needs of computational bioacoustics.

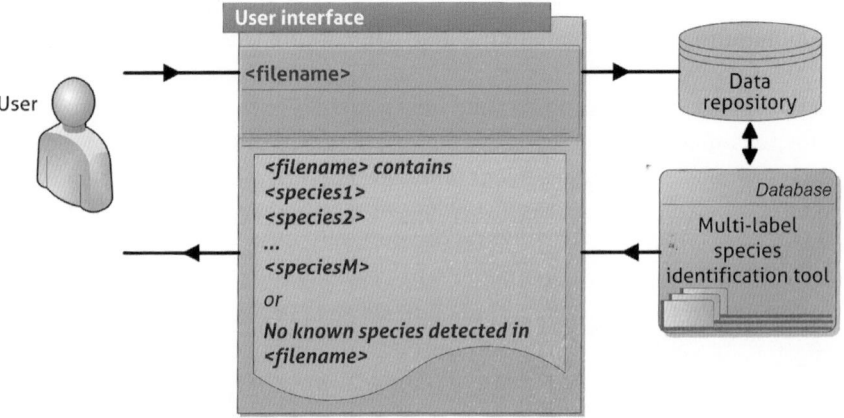

Figure 3.4: Conceptual diagram of the multi-label species identification task.

3.4 One-category recognition

In the *one-category recognition* task, we scrutinize long audio recordings aiming to select portions of audio where sounds from a broad category of species are present. Although there is just one category of interest, which has to be detected, this is a classic two-class decision problem, where the outcome is: *"Yes, sound events of* <category> *were found in* <filename>, *at the following start and end times*: <list>" or *"No, sounds of* <category> *were not found in* <filename>."* Depending on the selected pattern recognition approach, the one-category recognizer could be implemented with template-matching method or could be based on a single statistical model, which represents the category of interest. However, in practice the two-model approach is preferred as it copes better with noise and provides robust operation in real-world conditions. This advantage of the two-model approach is mainly due to the additional effort to model not only the category of interest but also the operational environment. Given that the model of the operational environment is comprehensive, this extra effort pays back with an improved discriminative power and robustness of the recognizer, especially when acoustic conditions are varying over time.

The one-category recognizer could serve as a gateway that passes for further processing only recordings containing sounds from a certain category of animals – only birds, only insects, only bats, and so on. As a result, the pattern recognition stage that follows will only process a subset of the continuous audio stream. For instance, let us consider a tandem of one-category recognizer and a species identification tool. Given that the one-category recognizer detects a sound event and its temporal boundaries, the subsequent species identification tool will process only that audio segment in order to find out whether it belongs to a known species for which there is prebuilt species-specific model. In this manner, the one-category recognizer contributes to the reduction of misclassification rates and computational demands compared to the case when the entire audio is processed with the species identification tool. Implicitly, here we assume that the one-category recognizer has lower computational demands than the species identification tool, which often holds true.

Another option is the one-category recognizer to be trained to recognize specific type of sound events, which are common among multiple species. For instance, certain alarm calls are common among several bird species – perhaps because purportedly some bird species have cross-species warnings for danger due to the approaching predator. Furthermore, some species have distinctive calls for the different types of danger, and therefore a category-specific recognizer tuned on a specific alarm call may assist (indirectly) in studies on the hunting behaviour or hunting tactics of some specific predator or on the specifics of cross-species communication. Predators are typically silent during hunting, and thus acoustic clues about the development of a hunting attempt can only come from the prey disturbed by the approaching predator or from other animals alarming about the episode of danger.

The *one-category recognition* task and the *one-species recognition* task (cf. Section 3.1) share a common machine learning approach; however, the models are built and used differently. A special case is the one-category recognition tool, which is tuned to discard silence and background noise, so that only acoustic events with energy above the noise floor are selected for further processing. Such a tool is referred to as an *acoustic activity detector* and is often used for reduction of the overall computational demands when implementing complex computational bioacoustic tools for *species identification* (Section 3.2) or *sound event-type recognition* (Section 3.8), and so on.

In the *one-category recognition* task (Figure 3.5), the user provides as input the name of the recording to be processed, that is, <filename>, and specifies which category of sounds should be selected for further processing. The outcomes of *one-category recognition* are the start and end timestamps of all recognized events, which specify the portion of audio kept for processing in subsequent processing stages. Therefore, the one-category recognizer acts as a gateway that passes for further processing selected portions of the audio and ignores the rest. Whenever the user input <filename> contains wildcards, and by that reason, there is more than one match in the data repository, each file is processed separately.

When the *one-category recognition* tool is implemented with statistical machine learning methods, two different data sets are needed in order to build the

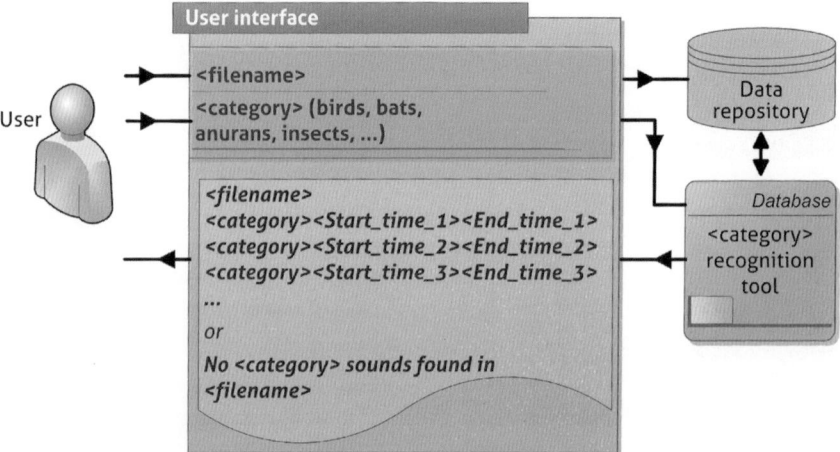

Figure 3.5: Conceptual diagram of the one-category recognition task.

two statistical models: the first one represents the acoustic properties of the selected category of sounds, and the second one the acoustic properties of the operational environment. Typically, the acoustic background model is created from large repository of audio recordings made in the operational environment, referred to as BL. These recordings need to be cleaned from any target sounds we aim to detect with the one-category recognition tool. The model which represents the target category of sound events is built from a TL, which contains a large number of examples, hopefully covering the intraclass variability of sound events falling in that category. The TL needs to be cleaned from any interferences and non-target sound events, so that the resulting target model is focused on the category of interest, and thus is able to provide good discriminative capability. The data cleaning typically involves selective manual editing of the audio in TL, and therefore is a time-consuming task. However, since the one-category recognizer needs just one TL the effort is manageable.

3.5 One-species recognition

The *one-species recognition* task aims to find out whether an audio recording contains sound event(s) originating from a certain predefined target species, and if *yes,* where exactly. For that purpose, the one-species recognizer has to detect the sound event and determine its temporal boundaries, that is, the timestamps of begin and end time in terms of relative or absolute time units. This is not a trivial task because the sound events of the target species often overlap with sound emissions of other species and with sounds of abiotic origin (wind, rain, vehicles, airplanes, etc.). The required functionality for timestamps estimation of sound events is the main difference when compared to the *one-species detection* task outlined in Section 3.1. In fact, this requirement results in a substantial difference in the design of technology because in the *one-species recognition* task we cannot simply make a binary decision *on per-file* basis (as we do in the *one-species detection* task). Instead, we need to scan the audio signal aiming to
1. differentiate between overlapping sound events,
2. discover the temporal boundaries of each sound event, and
3. decide whether it originates from the target species or not.

These requirements impose different uses of the species-specific model and additional data processing steps, which are not required in the *one-species detection* task.

In the most favourable use case, where sound emissions of different species do not overlap in time domain as there is some pause between subsequent events, we can carry out the *one-species recognition* by means of an acoustic activity detector followed by a one-species detector. In particular, the acoustic activity detector partitions a long audio recording to a sequence of sound events with known begin and end timestamps. Next, each audio segment is verified by the one-species detector that selects these belonging to the target species. Such a simplified species recognition scheme may work on audio recordings collected in a quiet controlled environment, for instance, in laboratory conditions with specimens in captivity. However, such a simplified design will not be appropriate for processing of soundscape recordings, where sound events originating from different species typically overlap in time and frequency range. This is mostly because of the lack of a mechanism for distinction of agglutinated sound events, which makes such a two-stage recognition scheme error prone. Specifically, such a two-stage recognition scheme is susceptible to errors due to merging of sound events of more than one species into a single audio segment, and therefore due to incorrect estimation of the temporal boundaries, or even due to a misclassification of the target species. These errors may significantly decrease the overall recognition accuracy and impede the discriminative power of the recognizer.

In the *one-species recognition* task (Figure 3.6), the user needs to provide two inputs: (i) the <filename> of a long recording to be processed and (ii) the <species> name for the target species of interest. It is assumed that there exists a prebuilt species-specific model for the specified <species>. After the audio processing is finished, the one-species recognizer outputs a list with the timestamps of begin and end time for each sound event associated with the target species, or alternatively, a notification that sound events of the target species were not found in recording <filename>.

The success of the one-species recognition process greatly depends on the specifics of the vocal repertoire of the target species and on the complexity of the acoustic environment in which the species was recorded. Species with

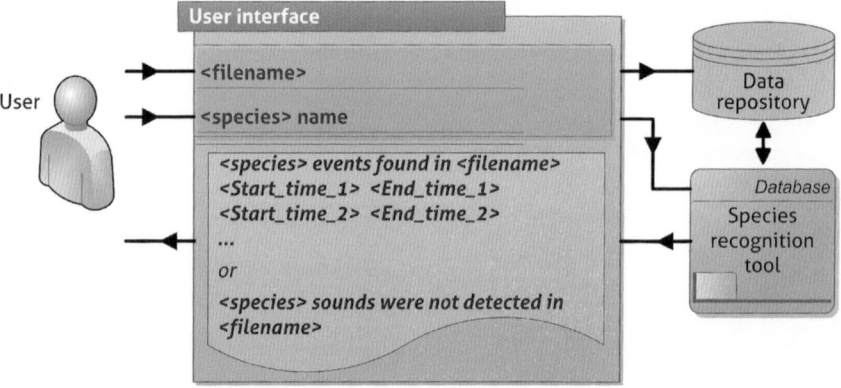

Figure 3.6: Conceptual diagram of the one-species recognition task.

well-localized harmonics in the frequency spectrum, simple repertoire, limited capacity for vocal interpretations, which are not inclined to learn new songs, are easier to model and recognize. The creation of representative acoustic libraries for such species would require fewer efforts when compared to species with complex repertoire. On the other hand, species with a repertoire of few dozens of calls or song types and with broad frequency spectrum are quite difficult to model and would require the availability of large comprehensive TL that provides multiple representative examples of each sound event type.

In general, most statistical methods for *one-species recognition* make use of the two-model approach, which builds models for the target species and for the acoustic environment. These models could be created independently of each other, or alternatively, the species-specific model could be derived through adaptation from the more general model of acoustic background. In Section 7.3, we discuss a particularly interesting scheme for the implementation of a one-species recognizer which makes use of two statistical models which are coupled: (i) model of the target species and (ii) model of the acoustic environment. The acoustic background model, which represents the acoustic properties of the environment, is built based on a purposely developed acoustic BL consisting of a large number of recordings collected in the operational environment. The BL is manually cleaned of sound events belonging to the target species. The acoustic model for the target species is derived from the acoustic background model through an adaptation process, which makes use of a species-specific TL. The TL is selectively edited manually in order to suppress sound emissions originating from other species. This particular way of model adaptation makes the two models tightly coupled, which improves the discriminative capability of the one-species recognizer.

Recent studies on the *one-species recognition* task carried out on recordings collected in field conditions were reported in Evan and Mellinger (1999), Brandes et al. (2006), Trifa et al. (2008a), Bardeli et al. (2010), Potamitis et al. (2014), and Ganchev et al. (2015).

3.6 Multispecies diarization

The *multispecies diarization* task usually operates on long audio recordings with the aim to create a chronological diary (and/or a chronogram) of all sound events, name their source (if known), and specify their timestamps. This task requires the implementation of the following processing steps:
1. Estimate the overall number of species that are acoustically active in each particular (soundscape) recording.
2. Segment the long audio recording to homogeneous partitions, each containing sounds of a single species or background noise.

3. Identify the species, to which these sound segments belong.
4. Determine the timestamps of begin and end for each sound event.
5. Create a chronological diary, which specifies the source (if known) and the timestamps of all sound events.

In the *multispecies diarization* task, prior knowledge about the number of species, which are acoustically active in each recording, is not a prerequisite, and in fact, such information is not readily available for soundscape recordings. However, when such prior information is available (e. g. after manual tagging), it facilitates the diarization process. This is because when the total number of species is unknown, it is more likely that errors which occur during the automated segmentation of audio will cause lumping together of vocalizations belonging to different species. The last is quite probable for species that share some common features, vocalize together or within a short interval one after another. Another type of error is due to incorrectly splitting the vocal repertoire of one species in two or more distinctive categories. Usually these two types of errors co-occur – some part of the vocalizations of one species is attributed to another one, and other sound events are considered as if they originate from a third species although they are form the first one. Therefore, this task is often implemented semi-automatically and iteratively with human expert in the loop of data analysis so that some errors are corrected at the initial iteration(s).

In addition to the species name, the *multispecies diarization* task may also provide the genus and family name. In such a way, the diary would incorporate useful information even when the sound events originate from species for which there is no prebuilt species-specific model, that is, unknown to the system (and in some cases unknown to science) species. In order to support such functionality, one has to build representative models for all genus and families of interest. These models are built based on acoustic libraries, each containing recordings of multiple species representative of the particular category.

Although the tasks discussed in Sections 3.3–3.5 can provide information about a specific category of sounds, the species name, and the timestamps of sound events, the *multispecies diarization* task is far more complex and challenging than any of the tasks discussed to this end. This is because the implementation of the *multispecies diarization* task requires a much broader functionality and more sophisticated tools, which collaborate and share information. For instance, the proper estimation of the number of acoustically active species in an audio file, or the accurate audio segmentation, will depend on an iterative procedure, which successively refines the estimations with respect to a certain predefined criterion. These criteria are closely related to the species identification and the timestamp determination functions. The iterative procedure might also

1. be aiming to discover all acoustic events of a certain species,
2. keep their timestamps and the species name in a diary,
3. clean the processed audio file from these segments in order for other weaker sound events to become more prominent, and
4. go back to (1) until certain task completion criteria are met.

In the *multispecies diarization* task, the user provides only the name of the long recording(s) <filename> to be analysed (Figure 3.7). After completion of the audio processing, the diarization tool outputs a diary and/or a chronogram, which contain information about the acoustically active species and the corresponding timestamps. The timestamps for the acoustic events could be presented in terms of relative time units, with respect to the beginning of each recording, or in terms of absolute time when the recording set-up permits this. Overlap between sound events originating from different species/specimens is rather common, so at times when the number of species is large the diary is more convenient to be represented graphically as a chronogram. The use of absolute time units has certain advantages as it allows the incorporation of prior knowledge and certain constraints when such information is available. Furthermore, the use of absolute time facilitates the subsequent interpretation of results, hence it is preferred when feasible.

When none of the known species for which there are prebuilt species-specific models is acoustically active in the specified <filename>, and none of the genus-specific and family-specific models was activated, the diary will contain a single entry – background noise – with timestamps set equal to the start and end times of the entire recording. In such a case, the user receives a notification that none of

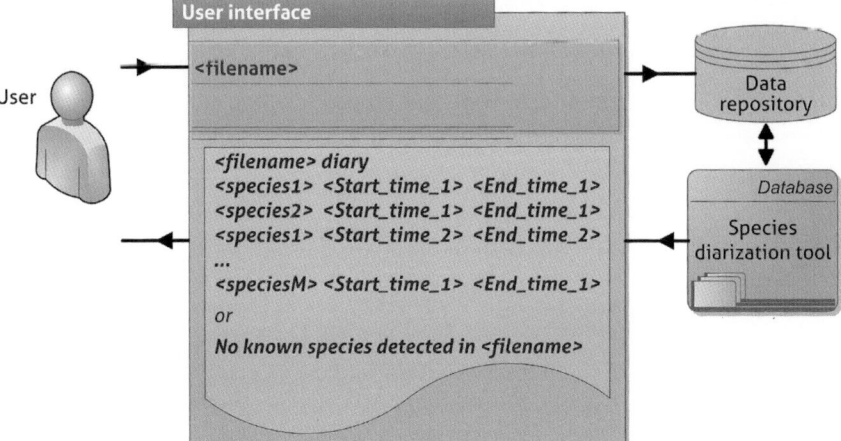

Figure 3.7: Conceptual diagram of the multispecies diarization task.

the known species was detected in the particular recording. Alternatively, one can relax the option for obtaining the species names, genus and family information, and only request the timestamps of homogeneous partitions of sounds. Such a relaxed process deviates from the diarization task; however, it could serve numerous other purposes, including providing valuable segmentation functionality in semi-automated annotation of soundscape recordings.

The species-specific models are built from a purposely developed species-specific TL similar to the *one-species detection* task (Section 3.1) and the *one-species recognition* task (Section 3.4). Alternatively, these models could be based on the *multi-label species identification* approach described in Section 3.3. A separate model, referred to as acoustic background model, is created to represent the acoustic conditions of the operational environment in the audio segments where prominent sound events are not present. This model is created from acoustic BL containing a large number of audio recordings collected in the operational environment that is characteristic of the particular application scenario.

At present, the *multispecies diarization* technology is not sufficiently mature for incorporation in real-world applications. However, we anticipate that in the near future it could serve in various content extraction schemes. For instance, it could be of particular help in biological census studies as it provides a computer-assisted support to tasks currently performed only manually and dependent on the involvement of highly qualified biologists and bioacousticians. Furthermore, the *multispecies diarization* can be seen as an important component of future automated systems for biodiversity monitoring. Finally, the *multispecies diarization* may become an important information source to an expert system, designed to provide decision support to management bodies of protected areas and natural reserves.

3.7 Localization and tracking of individuals

The *automated localization and tracking of individuals* task aims to determine the position (within a small area) and the trajectory of relocation over time for individual(s) of a certain species, based on their acoustic activity. The localization and tracking functionality is usually feasible only at short distances, in the range of metres. The actual range depends on the configuration of microphone array, the selected spacing between microphones, their sensitivity, and so on. Besides, the localization and tracking of species characterized with harmonic sounds is easier and more reliable than those with broad frequency spectrum sounds are. This is due to certain limitations of the contemporary methods for direction of sound arrival estimation.

3.7 Localization and tracking of individuals — 53

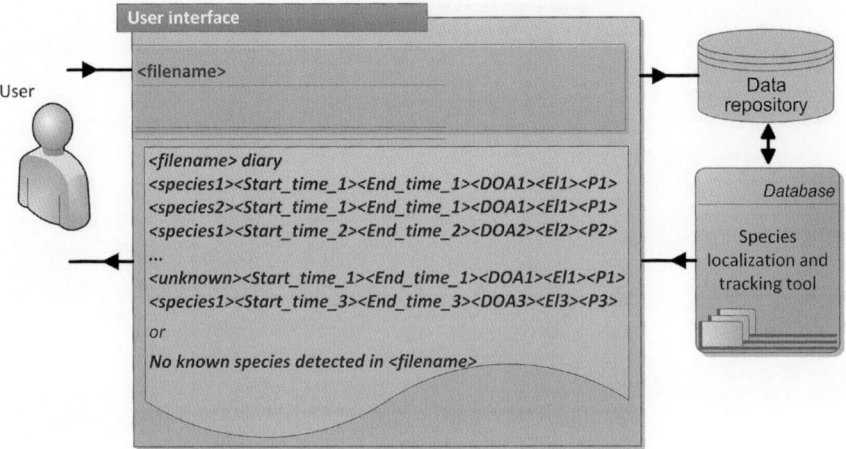

Figure 3.8: Conceptual diagram of the localization and tracking tool.

In the *multispecies localization and tracking* task, the user only provides the name of the audio recording to be processed <filename> (Figure 3.8).

After the audio processing is completed, the localization and tracking tool outputs a comprehensive report, which contains information about
1. the begin and end time of all detected sound events,
2. the species to which these belong,
3. the relative direction from which the sound arrived with respect to the orientation of the microphone array,
4. the height above the ground,
5. the intensity of sound source, and so on.

The species identification functionality requires prebuilt species-specific models, which could be created using any of the methods discussed in Sections 3.1–3.6. However, the estimation of the direction of arrival, elevation above ground, and intensity of sound might not need prior statistical modelling, and therefore these are feasible to determine the direction of any sound event. For instance, even for sound events that originate from unknown species (no species-specific model available) or are from abiotic origin.

In applications where the localization and tracking of individuals is required for a certain particular species, users have to specify both the species name <species> and the recording <filename> to be processed. In such cases, the tracking could be restricted only to individuals from the species of interest and the tracker will ignore all individuals from all other species, or alternatively could track all sound events but only the species of interest will be named and the others will be tagged as "unknown". However, here we need to note that the tracking of individuals from a specific species is implemented based only on their spatial localization over time and does not require the availability of prebuilt individual-specific acoustic model.

The multispecies localization and tracking task can be reduced to a number of single-species tracking tasks, which in turn is usually implemented with single-individual localization and tracking methods. Depending on the restrictions to computational complexity and time delay, the individual trackers can work independently of each other or can share information and/or the decision-making process.

The locality and tracking information well supports other tasks aiming at the recognition of activity patterns, behaviour recognition, abundance estimation, and population density assessment for a certain species. For that purpose, the localization and tracking tools typically work in tandem with the *one-species detection* or the *multispecies identification* tools (Trifa et al. 2007, 2008b).

At this point, we need to clarify that the present-day localization and tracking technology is not well suited for real-world environment, although research activities intensified in the recent years due to the wider use of wireless sensor networks and sound-enabled Internet of things (IoT) devices.

3.8 Sound event-type recognition

The task *sound event-type recognition* aims to distinguish among the different sounds produced by individuals of the same species and to estimate the temporal boundaries of each sound event. This includes all types of sounds among which calls, call series, songs, and so on, are typically present in the repertoire of a certain species of interest. By definition, this is a multi-class recognition problem, where an open-set identification of multiple types of sounds is carried within the temporal boundaries of each audio event. In this task, when there is a co-occurrence of several sound events we only recognize the dominant sound event, and therefore assign a single label to each audio chunk specified by the start and end temporal boundaries of that event.

The difficult multi-class identification problem could be reduced to multiple two-class detection problems. In such a way, each recognizer will be tuned to recognize one specific type of sound event and the individual sound event-type recognizers will have to be run on the same data set. However, in such a case the interpretation of recognition results obtained by the individual recognizers will require the implementation of a certain mechanism for resolving ambiguities.

The *sound event-type recognition* task is not a trivial task as for many species there is a significant variation in the sounds produced by populations inhabiting

3.8 Sound event-type recognition

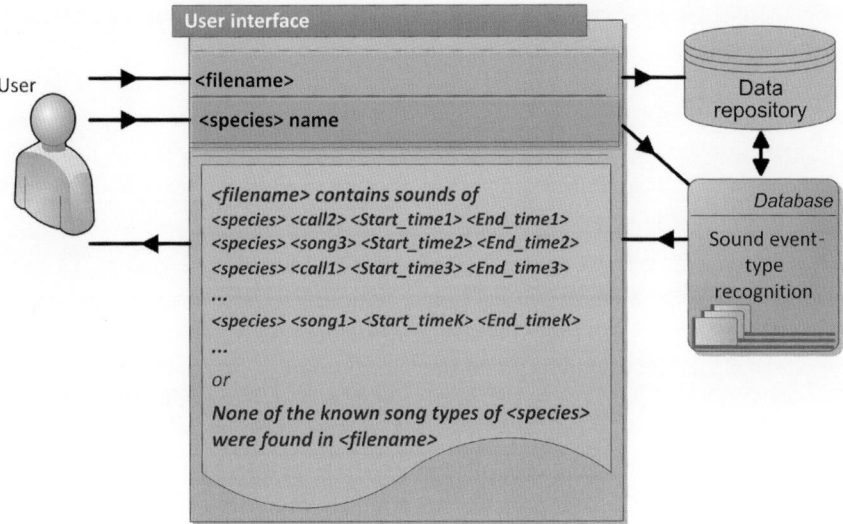

Figure 3.9: Conceptual diagram of the sound event-type recognizer.

different areas. Most often, this is because the different populations of these species adapted their vocal communication depending on the environmental conditions, and adjusted repertoire and timing depending on the context set by competing sound sources.

In the *sound event-type recognition* task, the user has to specify the species <species> of interest for which there are prebuilt sound event-type specific models, and the audio file name <filename> to be processed (Figure 3.9).

The outcome of this task is a list which includes information about the types of sound events in the repertoire of <species> that were detected in <filename>. When sound events specific to <species> were not detected in <filename>, the user is notified that none of the known song types for the species of interest were found in the specific recording. Whenever the user input <filename> contains wildcards, and by that reason, there is more than one match in the database repository, each file is processed separately. When users are interested to analyse the call and song types of more than one species, then the *sound event-type recognition* tool has to be run multiple times on the same set of audio files, each time specifying just one <species>. The last implies that multiple call and song models are available for a number of species.

The *sound event-type recognition* task imposes more stricter requirements to the species-specific TL when compared to the tasks discussed in Sections 3.1 and 3.2. This is because the creation of TL with different sound events (calls, call series, songs, etc.) for each species requires labelling to be carried out not only on species level but also on the level of the type of sound events.

These specifics require the involvement of expert bioacousticians who are well acquainted with all sound emissions of the species of interest and preferably with the context in which these sounds are produced. Besides, the TL has to be balanced in the sense that a sufficient amount of examples of each sound event type is present. The last is required in order to provide that the resultant statistical models are well tuned and are focused on the sound event types of interest. Under-represented types of sound events, for which the number of characteristic examples in the TL is not sufficient, will not be recognized properly and will cause high false-positive rates. Thus, the main difficulty for the creation of sound event-type recognition tools is the availability of representative and well-balanced TL, especially for species that have a large number of distinct sound emissions. For instance, some species may have a repertoire of over 30 or 40 different calls, call series, and songs, which makes the creation of a comprehensive TL time-consuming, expensive, and, in most cases, prohibitive for practical applications.

For species that have a large number of distinctive sound emissions, it might be more convenient to make use of multiple recognizers, each specialized for a single song type. These recognizers operate on a common input. In such a way, one can avoid the need to retrain all models in the recognizer when a new song type is added. Each specific recognizer could be built independently of the other or could be derived from a general acoustic model (cf. Sections 3.2 and 7.3).

The *sound event-type recognition* only makes sense in field studies associated with the recognition of activity patterns and behaviours, and therefore technology has to cope with competing sound emissions from multiple species and non-stationary noisy conditions. The last requires proper modelling of the acoustic environment in the specific habitat. State-of-the-art methods for *sound event-type recognition* rely on statistical models, which are representative of the operational conditions of the recognizer. These models are built based on an acoustic BL, which contains a large amount of recordings made in the specific operational conditions. The BL is cleaned from any sound emissions of the species of interest for which the call, call series, or songs need to be recognized. However, when we need to build *sound event-type recognizers* for multiple species, the creation of individual BL for each recognizer is not practical, neither the cleaning of the BL from all target species. Due to the large size, BL cleaning might require a significant investment of time. The last enforces the use of a simple BL that is common to all species and that does not contain too many loud sound events but weak sounds close to the noise floor.

The *sound event-type recognition* contributes information to higher-level analysis tasks, such as abundance assessment, activity patterns recognition, and behaviour recognition studies. The information about specific sound event is associated with other information and subsequently is interpreted depending on the available contextual information.

At present, the automated *sound event-type recognition* is not used frequently due to the lack of comprehensive TL and BL and the prohibitive effort needed for the creation of acoustic libraries. One opportunity for relieving these difficulties would be to foster research towards combining the advantages of template-matching pattern recognition methods (cf. Anderson et al. 1996) and statistical methods (cf. Trifa et al. 2008a, 2008b) for song-type recognition through multi-stage modelling and recognition approaches. Finally, the *sound event-type recognition* bears similarity with speech recognition and thus, given the availability of comprehensive TLs, it could benefit from applying state-of-the-art methods and well-conceived technological solutions already developed for speech and speaker recognition applications.

3.9 Abundance assessment

The *abundance assessment* task aims to estimate the number of individuals of a certain species within a given area only by means of the observed acoustic activity. This estimation is most often based on indirect clues and some species-specific assumptions. For instance, an estimation of the number of individuals in an area could be based on evidence of acoustic activity measured with *one-species recognition* tool (Section 3.4), counting the number of species-specific sound events per unit time. In that case, it will be important to distinguish between sound events produced by single individuals, pairs, and chorus (if applicable for that species) by means of species-specific *sound event-type recognizer* (Section 3.8). On the other hand, distinguishing among individuals could be based on spatio-temporal localization set-ups similar to those in the *multispecies localization and tracking* task (Section 3.7) or even on direct methods for acoustic differentiation between individuals based on certain traits of their sound emissions. The last is feasible only for a small subset of species, and considering the challenges imposed by real-world conditions is beyond the reach of present-day technology.

There are recent studies reporting on acoustic recognition of mammals (elephants, whales, etc.) based on methods developed for the purpose of speaker recognition, that is, methods developed for human speech. However, these methods require the availability of tagged audio recordings from the specific individuals of interest – in order to create individual-specific models – and from a large number of other individuals in order to create a general model representing the rest of the population (Kinnunen and Li 2009). Then scoring the input against the individual-specific model and the general model, we can recognize whether a certain individual is acoustically active or not in a recording. This approach might be applicable for tracking the acoustic activity of a small group of known individuals in the area of study; however, it is not applicable for the needs of *abundance assessment* task. Obviously, it is not practically possible to record all individuals of a certain species living in the study area and then create models and a recognizer in order to count how many of these individuals are acoustically active during a certain period of time.

Thus, although on a first glance there seems to be a connection and common traits between the speaker recognition for humans and the recognition of individuals of a certain species – this does not help the *abundance assessment* task. In the *abundance assessment* task, we aim to distinguish among acoustic emissions from different individuals only by being able to count how many different individuals are acoustically active in the area of study. This is extremely challenging to be achieved only through audio.

Abundance assessment could provide raw data to subsequent higher-level data analysis and interpretation tasks, such as the population size and density estimation. These are currently performed only by human experts and therefore are of limited scope, time-consuming, and expensive. Advancements in the area of computational bioacoustics are expected to facilitate the *abundance assessment* methods and the related data analysis and interpretation tasks.

3.10 Clustering of sound events

The *clustering of sound events* task aims to group together sound events based on predefined similarity criterion or criteria. This functionality could facilitate any efforts towards automating sound event-type or species-level annotations, a process which is currently performed mostly manually by highly qualified bioacousticians. The automated grouping of audio events could be implemented based on
1. a specified etalon and certain similarity criteria or
2. only on predefined criteria applied on a certain set of audio descriptors.

In the case when exemplary etalon sound event is provided, it serves as a seed around which other sound events are congregated based on the selected similarity criteria. This is extremely useful in semi-automated annotation of audio recordings, for example, when an expert bioacoustician selects an event of interest, and then the automated tool finds all other occurrences of this sound event in the recording.

When the grouping of sounds is based only on applying the predefined criteria on the audio descriptors, then the groups might not have meaningful names. The most frequently observed sound types would be congregated in groups based on the selected similarity metrics. Even if these groups do not have meaningful names, they could help in uncovering the underlying structure of sound emissions and might be useful in achieving a better understanding about (i) the sound activity patterns of certain species or (ii) the similarity of calls or syllables produced by different species. The last helps to improve our understanding about the potential sources of false alarms (in the one-species detection and the one-species recognition tasks), and the risks of misclassification (in the multispecies identification task).

Furthermore, the *clustering of sound events* can support information retrieval tasks as it allows the implementation of query-by-feature or query-by-example functionality based on selected similarity criteria or features (Figure 3.10). In this manner, one can implement search for a specific sound entity in a recording or in an audio repository. The entity of interest could be a specific call type, a call sequence, a song or a subunit of song, a sound event, and so on, or may not have a meaningful name.

In the case of certain less-complex sound events, it is possible to implement the query-by-example functionality, that is, to define a query in the format "*search for this chunk of sound in this audio recording.*" For that purpose, we can use direct template-matching techniques, such as those reported in Anderson et al. (1996), Fodor (2013), and Stowell and Plumbley (2014), or memory-based classifiers such as the *k*-nearest neighbours, probabilistic neural network, and so

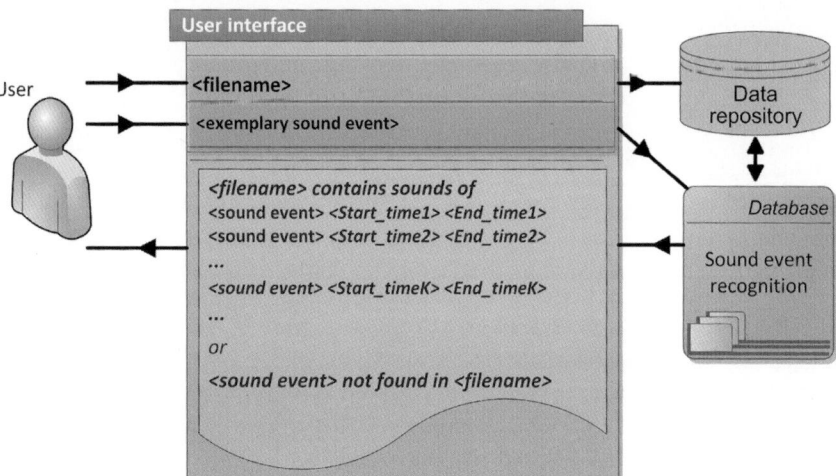

Figure 3.10: Conceptual diagram of the query-by-example search.

on. If multiple instances are available for the sound event of interest, one could utilize statistical modelling techniques such as the Gaussian mixture models or the hidden Markov models. We provide more details on the various classification approaches in Chapter 7.

The chunk of sound to be searched for is usually selected manually through a graphic user interface (e. g. ARBIMON-Acoustic web application,[4] Song Scope Software[5]), where the user can draw a box over the audio spectrogram of a recording. The box specifies the temporal boundaries and the frequency range of the audio event of interest and will ideally not include competing sound events. For that purpose, selective manual filtering of the box content might need to be carried out in order to eliminate interferences. The presence of interferences might not prevent the search from successful outcome; however, discarding these is recommended for obtaining an optimal outcome of the search process and reducing the chance that less relevant chucks of audio sneak in the results list. When the sound event of interest originates from a recording made in laboratory conditions, and thus is noise-free, a constraining box is not required because the signal processing algorithm automatically discovers the temporal boundaries and the frequency range of sound events. In such a case, the audio chunk <exemplary sound event> could be uploaded directly to the search engine.

3.11 Biodiversity assessment indices

The use of automated recording devices allows researchers to implement continuous audio recording simultaneously at multiple locations and microsites over long periods of time – in the range of months and years. Often within a research project, for example, the ARBIMON project,[6] the AmiBio project,[7] the INAU project 3.14,[8] SABIOD project,[9] and other long-term projects, researchers collect hundreds of terabytes of recordings, amounting to many thousands of hours of audio. However, the time required for the manual processing and analysis of the acoustic emissions captured in these recordings is much longer than the total duration of recordings. This processing is particularly time demanding when analysis and interpretation activities are performed manually with traditional human expert-based bioacoustic methods. Moreover as discussed in Sections 3.2–3.10,

---

4 ARBIMON-Acoustic web application, http://www.sieve-analytics.com/#!arbimon/cjg9
5 Wildlife Acoustics Ltd, http://www.wildlifeacoustics.com/products/song-scope-overview/
6 The ARBIMON project, https://www.sieve-analytics.com/arbimon
7 The AmiBio project, http://www.amibio-project.eu/
8 The INAU project 3.14, http://www.inau.org.br/laboratorios/?LaboratorioCod=12&ProjetoCod=94
9 The SABIOD project, http://sabiod.univ-tln.fr/

automated audio analysis methods typically depend on the availability of TLs, which are annotated on the level of species or sound events. The creation of such libraries is a time-consuming task and requires expert knowledge, which makes the process expensive and even prohibitive when the automated recognition of multiple species is required.

In order to alleviate these difficulties, recent studies investigated an alternative approach for soundscapes ecology analysis (Pijanowski et al. 2011a, 2011b) – based on acoustic indices calculated at different timescales (Sueur et al. 2008; Pieretti et al. 2011; Depraetere et al. 2012; Gasc et al. 2013; Farina 2014; Lellouch et al. 2014; Towsey et al. 2014; Eldridge et al. 2016). Such acoustic indices provide integral information about the overall acoustic activity over time and supposedly demonstrate a good correlation with the species richness in the study area.

A comprehensive study of 14 acoustic indices is available in Towsey et al. (2014). These indices fall into two broad categories: (i) standard indices for assessment of audio quality and (ii) acoustic indices specially designed for measuring the informative content of recording, or of sound features characteristic to the sound-emitting species of interest. The standard indices evaluated there were *signal-to-noise ratio*, *signal amplitude*, and *background noise floor*. The second group of indices included
– *Acoustic activity* – the fraction of audio frames for which the signal envelope is above the background noise by at least 3 dB
– *Count of acoustic events* – the number of times the signal amplitude crosses the 3 dB level (above the noise floor) upwards within a certain time interval
– *Average duration of acoustic events* in milliseconds
– *Entropy of the signal envelope (temporal entropy)*
– *Acoustic complexity index*
– *Mid-band activity*
– *Entropy of the average spectrum (spectral entropy)*
– *Entropy of spectral maxima*
– *Entropy of the spectral variance*
– *Spectral diversity* – the number of distinct spectral clusters in a 1-min recording
– *Spectral persistence* – an index that reveals the degree of similarity between successive audio frames

Certain combinations of these acoustic indices were reported useful for assessing the species richness in natural environments (Towsey et al. 2014), which is quite important for ecologists engaged with monitoring of terrestrial ecosystems. A critical overview of acoustic indices and in-depth discussion on their fundamental constraints is available in Eldridge et al. (2016).

3.12 Practical difficulties related to real-world operation

Besides the purely scientific curiosity, which was particularly distinctive for the early years of bioacoustics, nowadays there exists an attitude towards the practical use of automated species recognition technology. This is especially true for services that facilitate the data annotation, analysis, and interpretation, as described in Sections 3.2–3.10. A number of commercial products and services, which incorporate species recognition components (e. g. Wildlife Acoustics,[10] Sieve Analytics[11]), were launched on the market and have already started to attract significant attention.

Besides the fact that the acoustic species recognition and sound event recognition technology has not yet reached sufficient maturity to meet the challenges of robust operation on real-world recordings, some market niches already offer good opportunities for successful commercial applications. Such a niche is offered by services that recognize bird sounds and allow online identification of birds on mobile devices. These services are usually based on the client–server architecture, where the sound is recorded by means of ordinary smart phone or another mobile device and the recognition process is carried out on a remote server (nowadays easily implemented as a cloud-based service). On the other hand, with the present-day technology at hand, the problems due to distortions of the audio signal introduced during its acquisition and transmission through the communication channel, and interferences from the environment, can be addressed in a satisfactory manner. Still, the practical challenges related to the species recognition process are many. The most prominent among these are
- requirements for high performance, robustness, and flexibility;
- availability of TLs or bootstrap data for initial training;
- need of model adaptation to different acoustic conditions or new area;
- appropriate decision strategy according to the operational set-up;
- dependence on species-specific prior knowledge about the behaviour and performance of species, and so on.

Due to these particularities, the brute-force approach would not allow achieving optimal performance. In that sense, the species recognition process cannot be merely reduced to a machine learning problem requiring the extraction of some audio features, statistical modelling, and subsequent decision-making. Such simplification would entirely overlook the species-specific factors and the difficulties related to operational environment. Therefore, the efforts spent in studying

---

[10] Wildlife Acoustics Ltd, http://www.wildlifeacoustics.com/
[11] Sieve Analytics Ltd, https://www.sieve-analytics.com/

the species peculiarities and incorporating species-specific knowledge in the recognizer usually will pay back with an improved performance.

Furthermore, the real-world deployment and exploitation of species recognition technology involves multiple application-specific trade-offs in terms of
- the availability of appropriate audio libraries for model training, fine-tuning, and validation;
- logistics and maintenance cost for periodic servicing of equipment;
- time for training/adaptation of the models and adjustment of system parameters according to the operational conditions;
- robustness of operation under varying noise conditions in the presence of strong interferences and competing sounds from multiple sources;
- achieving a reliable performance of equipment in adverse climatic conditions;
- adaptability to changing requirements and reprogramming of functionality;
- overall data communication and data processing delays, and so on.

The successful establishment and exploitation of an automated technology is conditioned largely on the manner the above-mentioned challenges are addressed. Obviously, it all depends on the proper organization of the cross-disciplinary teamwork, purposeful joint effort, and the talent, professionalism, and devotion of all team members. In the following chapters, we consciously constrain our discussion to the technology development perspective. Specifically, in Chapter 4 we focus on the design of appropriate audio libraries. These audio libraries are a prerequisite for the creation of statistical models, fine-tuning of decision-making process, and performance validation, which are essential steps in all research and technology development tasks discussed in Sections 3.2–3.10. In this regard, in Chapter 4, we discuss the purpose of various audio libraries and describe their contents, usage, and preparation.

4 Acoustic libraries

Introduction

As discussed in Sections 3.1–3.9, computational bioacoustic research and technology development activities depend on the availability of manually selected and, in some cases, duly annotated audio data sets. These data sets serve different purposes, among which are the
1. creation of statistical models,
2. fine-tuning of various adjustable parameters of the automated technology,
3. evaluation of the overall system performance, and so on.

In Sections 4.1–4.4, we discuss in detail the content, design, and specifics of these audio data sets, which we refer to as *acoustic libraries*. In order to facilitate the exposition, we briefly mention the purpose and practical use of each of these acoustic libraries. Specifically, here we distinguish among (i) *training libraries* (TLs), (ii) an *acoustic background library (BL)*, (iii) a *validation library (VL)*, and (iv) an *evaluation library (EL)*, which are described as follows:
- TLs are used for the creation of category-specific models, which represent the target categories of sound events. Here, we mainly refer to category-specific *acoustic models*. However, in the case of species with complex sound repertoire, TLs (and in particular their annotation) also constitute an important source of information for the establishment of semantic rules and grammars. The combined use of acoustic models and grammars permits an enhancement of the overall recognition accuracy of automated recognizers. TLs usually consist of a large amount of annotated (or tagged) segments of audio recordings. Depending on the particular recognition task, the annotation might be implemented on the level of song type, sound event, species, genus, families, and so on. It is worth mentioning that in the process of TL creation the recordings are usually hand-edited for the elimination of sounds coinciding in time (or overlapping in frequency range) with the target category. The last is considered necessary because the majority of the statistical machine learning approaches (cf.Chapter 7) typically assume that one category-specific TL is available for each target sound category. Therefore, most frequently, the corresponding category-specific models are created without additional effort for vigilant cleaning of audio recordings. Thus, a TL is expected to contain only sound events that are representative of the target category which need to be modelled and should not include other types of sound events unless these have a negligible amplitude.

- *An acoustic* BL is used for the creation of a general acoustic background model, also referred to as universal background model or a world model, which is representative of the acoustic environment where the specific technological tool will be deployed to operate. In generative classifiers,[1] the resultant acoustic background model is typically used as a reference for the normalization of target model scores. Another use of the acoustic background model is for approximation of the alternative hypothesis in acoustic detectors (cf. Section 7.2). By that reason, the BL typically comprises of multiple soundscape recordings, which are free of sounds corresponding to the target categories aimed to be modelled trough the TLs. When such recordings are not readily available, or if their cleanness is not fully guaranteed, the BL needs to be cleaned manually or semi-automatically for attaining a better performance. An important difference with respect to the above-mentioned TLs is that the BL does not need to be tagged or annotated for specific acoustic events. However, in order to provide a fair and unbiased representation of the acoustic conditions in a real-world operational environment, the BL usually needs to be many times bigger than the TLs. The last is because in a real-world set-up the acoustic environment is time varying, and its characteristics are conditioned on a multitude of uncontrollable factors (wind, rain, acoustic activity of non-target species, human activity, seasonal changes in vegetation, etc.).
- *A* VL is used in the process of technology development, primarily for fine-tuning the adjustable parameters (and thus the performance) of a certain technological component, or of an entire system. In this regard, the fine-tuning might aim at (i) adjusting the settings of a noise reduction front-end to the operational acoustic environment in which technology will be deployed, (ii) calibration of certain parameters with the help of carefully selected etalon signals with known parameters, (iii) tuning some adjustable parameters of the modelling or decision-making process, and so on. Because of that, the VL usually consists of purposely selected portions of soundscapes, or entire soundscapes, which are representative of the real-world operational conditions. The selected audio is then annotated for the target sound events of interest. The labels and timestamps defined during annotation are referred to as the *ground true* labels of the data set. These labels are required in order to measure the system performance in the process of fine-tuning and assessment.

---

1 A thorough discussion of the principles and specifics of generative and discriminative classifiers is available in Jebara (2004).

– *An* EL provides the means for a realistic performance assessment of an acoustic recognizer in an experimental set-up, which approximates as close as possible to the real-world operational conditions. Therefore, the proper selection of soundscape recordings for the EL is crucial for obtaining an unbiased evaluation of recognition performance. The EL typically consists of entire soundscapes collected in the same operational environment where technology will operate. These soundscapes are unprocessed and when possible kept in their original format, frequency bandwidth, sample resolution, and so on, that is, unaltered in any way. However, the audio needs to be manually annotated for the target sound events of the categories of interest, so that the ground true labels and timestamps are known.

In order to illustrate the practical use of these acoustic libraries, let us consider the signal processing workflow shown in Figure 4.1, which is common to generative audio processing approaches (cf. Chapter 7).

In brief, we build, fine-tune, and evaluate an acoustic recognizer by making use of a *TL*, *VL*, and *EL*. This recognizer also makes use of an *acoustics BL*, which is needed for the creation of a general acoustic background model. In the specific design, the acoustic background model is considered representative of

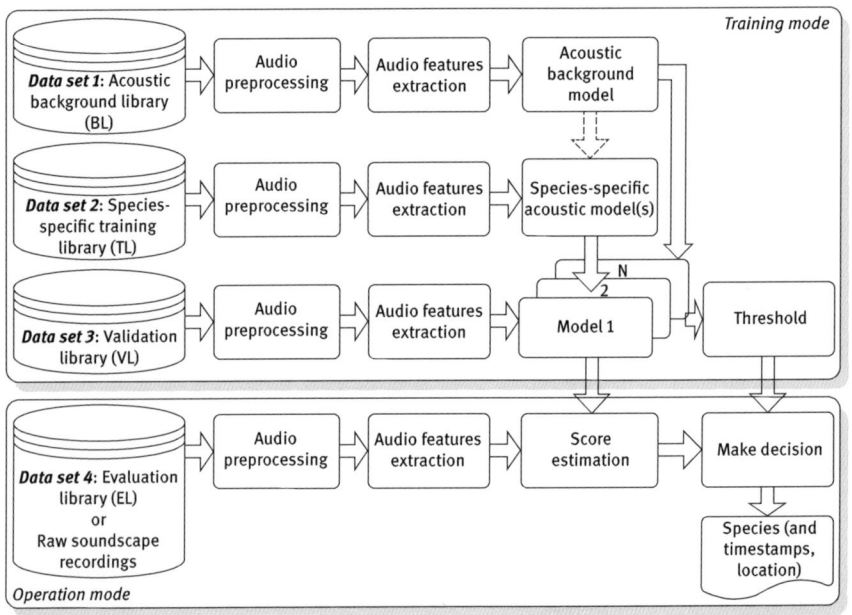

Figure 4.1: The role of various acoustic libraries used for the purpose of technology development.

the acoustic environment where the recognizer will operate. When the acoustic background model is used to represent the alternative hypothesis, we make the assumption that it represents everything except the target species, and therefore the target model(s) and the background model provide an exhaustive coverage of all possibilities.

As shown in Figure 4.1, in some generative machine learning methods it is feasible to derive the category-specific target models from the general acoustic background model (cf. Section 7.2). In other methods, the target models are created independently of the acoustic background model and are based only on the available category-specific TLs. The choice of model creation strategy depends on the specifics of each application; however to some extent, it is conditioned on the amount and the quality of representative data in the TLs. When the TL for some of the target categories is small, model derivation from the acoustic background model is advantageous when compared to the independent model created from scratch.

In the above-mentioned generative machine learning methods (Section 7.2), the acoustic recognition task is implemented through a comparison of scores, which are obtained after the evaluation of an unlabelled input against the pre-built models for the target categories and a general model that represents the alternative hypothesis. As a substitute to this mainstream pattern recognition approach, one might want to build a cohort of acoustic models created for categories of sounds that bear some similarity to the target category of interest. In the latter case, the final decision for an unlabelled input will depend on the scores computed through the

1. category-specific model of the target category,
2. respective category-specific cohort models, and
3. acoustic background model if it is present.

In both cases, the use of an acoustic background model helps in compensating the influence of time-varying operational conditions. Furthermore in detectors and classifiers, the use of cohort models helps in sharpening the decision boundary between the specific target category of interest and other categories that bear similarities to it. Therefore, both the acoustic background model and the set of cohort models contribute towards reducing the misclassification rates and are especially important for decreasing the rate of false-positive decisions. Such reduction of error rates is feasible only when the cohort models and the acoustic background model are created from carefully prepared acoustic libraries, which are representative of all categories of sounds.

As shown in Figure 4.1, once the performance of an acoustic recognizer is fine-tuned with the help of a *VL*, the training phase is completed. Next, the recognition performance is evaluated with the help of an *EL*, and if found satisfactory, we can proceed with deployment of technology for operation in the preselected location. The last holds true both for online acoustic recognizers that operate with real-time streams of data and for tools that implement offline processing of audio recordings.

In Sections 4.1–4.4, we provide details of the purpose, significance, and specifics of each of the four types of acoustic libraries, describe an example workflow for their creation, and provide a raw assessment of the overall effort needed. In addition, we comment on the main limitations and challenges related to data collection and preparation as these depend on the labour of highly qualified bioacousticians. The last is important because in many cases the amount of efforts needed for acquisition, preparation, and annotation of recordings is prohibitive, and without an efficient technological support the creation of suitable acoustic libraries is not feasible.

While the organization and content of acoustic libraries are well defined, the workflow of acoustic libraries creation discussed here is not exclusive and thus may differ depending on the requirements of each particular application. Therefore, aiming at illustrating the challenges, the scale of effort, and explaining the main differences in the creation of TL, BL, VL, and EL, we outline relevant workflows in a few general steps. In any way, we do not assert exhaustiveness of exposition as there could be various other approaches to acoustic library creation and these could differ in the overall workflow, and in the particular signal processing steps.

4.1 The training library

4.1.1 TL organization

Here, we define the notion of acoustic TL as a collection of purposely selected, edited, and labelled portions of audio recordings, which are archetypal for a certain predefined category of sounds. The purpose of TLs is to provide the raw data required for the creation of category-specific statistical models. A TL is the source of information for the creation of an acoustic model, a syntax model, or both. Thus, each TL has to contain multiple characteristic examples of the specific category of interest so that statistical machine learning methods are given the opportunity to model consistently the properties of data distribution or to determine accurately the decision boundaries among different categories. Furthermore, the generative statistical methods also aim to model the intraclass variability within each category, so that availability of comprehensive and representative TL is highly desirable. Both groups of methods – discriminative and generative – are discussed in Jebara (2004). In Chapter 7, we briefly mention some widely used classifiers and provide examples of their practical use in some computational bioacoustic tasks.

The nominal size of a TL required for the creation of a robust acoustic model depends on the inherent variability of sounds within each category that needs to be modelled. For instance, a species-specific TL for the one-species detection task (cf. Section 3.1) would greatly depend on the vocal repertoire of the target species. Species with a large repertoire of calls and songs would require a much bigger TL when compared to species that use only a few different call types. However, an unambiguous general rule that could provide an unbiased estimate of the required size of a TL for each specific case is not readily available. This is because the size of TLs depends on many other uncontrolled factors, among which are the complexity and dynamics of acoustic environments. Other factors that affect the required size of TLs are the number, and the statistical properties of audio descriptors, the model order, and model training method, and so on.

Nevertheless, based on empirical evidence and experience with acoustic modelling of various insects, amphibians, avian, and mammal species, we can still make a well-founded guesstimate that a TL containing annotated audio in the range of few minutes to few tens of minutes, for each category of interest, is a reasonable one. Here, we refer to the cumulative length of all annotated target segments of sound. However, when the target sound emissions are with low variability much less data are required for the creation of a consistent statistical model. For instance, this is the case for many insects, frogs, and some birds.

In all cases, the quality of statistical models is conditioned on the assumption that all annotations of audio segments are representative of the categories of interest. When irrelevant sound events make their way into the TLs, this distorts the consistency of derived statistical models in the sense that models are less focused on the category of interests for which these were created. One consequence of a less-focused model is that the risk of increased false-positive error rate becomes more pronounced.

Furthermore, the TL size required for the creation of a robust statistical acoustic model depends on
1. the choice of audio parameterization method;
2. the dimensionality of descriptors, viz. the feature vector size;
3. the assumptions made about the distribution of these descriptors (during post-processing of descriptors and during the creation of statistical models);
4. any use of prior knowledge about the model structure for the specific category;
5. the choice of machine learning method;
6. the implicit and explicit assumptions embedded in the selected methods for model training and refining, that is, the methods used for tuning the adjustable parameters of the model, and so on.

For instance, an audio parameterization method that takes advantage of prior knowledge about the sound production mechanism of species, or that accounts for the species-specific properties of sound emissions, is expected to be more successful when compared to any general audio descriptors estimation method[2] borrowed from speech processing or another loosely related audio processing application. It is quite important that the properties of signal descriptors, that is, their statistical distribution, variability, correlation, and so on, be well understood beforehand post-processing and modelling. The last is important because it facilitates careful planning of post-processing steps and permits subsequent reduction of the size of multidimensional feature vectors.

The dimensionality of feature vectors directly influences the minimum required size of TLs needed for the creation of satisfactory models. In the general case of generative pattern recognition methods, in neural networks, and other groups of methods, larger feature vectors impose the need of larger TLs. However, some discriminative pattern recognition methods such as those based on support vector machines (SVMs), decision trees, and so on, provide satisfactory results even with a smaller number of carefully selected training samples. This is because discriminative classification methods do not need to model the intraclass variability for each category of sounds but only aim to distinguish between two or more categories. However, many instance-based classification methods, such as probabilistic neural network, nearest neighbour classifier, and so on, can also provide satisfactory results with just few carefully selected training samples.

Prior knowledge about the species-specific sound production mechanism and knowledge about the properties of sound events permit the accurate selection of the model structure and order. The last is important for the efficient use to the best advantage of the available TLs and for reducing the chance of overfitting the model on the particular training data set. When the model order is unreasonably high, it starts to capture non-informative variability that reflects the influence of noise and other factors irrelevant to the specific task. The last is counterproductive with respect to modelling and worsens the generalization properties of the model, and thus the recognition performance for the category of interest.

---

[2] For instance, the Mel-Frequency Cepstral Coefficients (MFCC), Linear Frequency Cepstral Coefficients (LFCC), and other well-known audio descriptors are commonly used in the automated sound event recognition, species recognition, and other bioacoustic tasks. This is motivated with arguments that (i) these are good descriptors of the spectral envelope, (ii) are inspired by the human auditory system, and (iii) because of the perception that these are sufficiently good and do the job well. However, the correct question here would be whether a set of audio descriptors is optimal for the specific sound production mechanism. A detailed description of numerous general-purpose methods for speech and audio parameterization is available in Ganchev (2011).

Although to a smaller degree other implicit or explicit assumptions embedded in the model creation also indirectly influence the required size of TLs. Among these are the
1. selection of method for model training,
2. choice of metrics embedded in the cost function,
3. cost function design,
4. stop criteria of training process, and so on.

The TL design can be finalized only after we
1. define the sound categories of interest that need to be modelled;
2. have a clear idea about the statistical properties, intraclass variability, and diversity of sound events in each category; and
3. deduct the minimum acceptable size of the TL.

Once the design is finalized, TLs are implemented based on a specific workflow through manual editing of a preselected set of recordings or in a semi-automated fashion.

4.1.2 Creation of training libraries

As already highlighted, the availability of high-quality category-specific TLs is essential to the creation of robust acoustic models. For instance, when we consider the specific requirements of the one-species detection (Section 3.1), the species identification (Section 3.2), and the one-species recognition (Section 3.5) tasks, the TL would ideally contain recordings with acoustic emissions from a single species per file or at least the sounds originating from the species of interest have to be dominant. The last would mean that only the acoustic emissions of the target species are with amplitude high above the noise floor.

When audio is collected in a controlled set-up – for instance, when the species of interest are recorded manually with a parabolic microphone, or when recordings are made in laboratory set-up with animals in captivity – this is somehow possible to handle. However, in unattended real-world recording set-up, which aims at capturing soundscapes with omnidirectional microphones attached to autonomous recorders, it is impossible to control the quality of sound recording at all times. In this regard, it is impossible to guarantee that (i) the target species will be close enough to the microphone, so that the loudness of their sound emission will be high above the noise floor, or that (ii) concurring in time-competing sound events of other origin will not be present.

Both conditions are difficult to guarantee, and therefore when TLs are based on soundscape recordings the audio files have to be preprocessed manually or semi-automatically. By that reason, the preparation of TLs often requires long-term devotion by experienced bioacousticians, who need to listen to many recordings and inspect their audio spectrograms in order to select suitable portions according to some predefined specifications. These portions of audio are afterwards manually edited in order to eliminate overlapping sound events of competing species. Each portion is then either tagged on a file level or annotated on a sound event level in order to determine the temporal boundaries of all sound events belonging to the species of interest. In the following, we discuss manual and semi-automated implementation of the TLs' creation process.

4.1.3 Manual creation of TLs

Assuming that TLs need to be created based on soundscape recordings, let us consider a four-step process, which illustrates the basic workflow (Figure 4.2). In brief, given the availability of unprocessed soundscape recordings, we first prescreen the audio, select portions (*chunks, snippets*) which contain sound events of the target species, eliminate non-target sounds, and carry out an annotation in order to determine the temporal boundaries of sound events, or alternatively simply tag each chunk of audio.

First, not all soundscape recordings are guaranteed to contain acoustic events originating from the species of interests, and therefore, one may need to prescreen large number of audio files until the necessary amount of examples is acquired. Furthermore, the average number of sound events per unit time depends on season, location, daytime, weather conditions, activity patterns of species, and so on, and therefore, might vary greatly among soundscape recordings. Thus, many files might need to be screened until we select a suitable subset of soundscape recordings.

Eventually, the excellence of a TL is assessed in terms of its representativeness, comprehensiveness, accuracy of labelling, sound quality, purity from

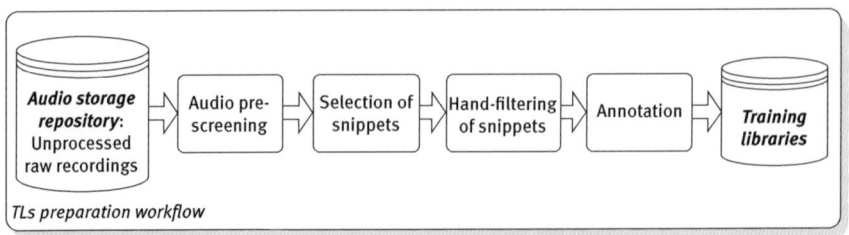

Figure 4.2: Diagram of a plausible workflow for manual creation of TLs.

artefacts, and so on. All these factors depend on the involvement of human experts at all stages of audio prescreening, selection, editing, filtering, and annotation. This requires the involvement of an experienced biologist, or a bioacoustician, in the audio prescreening and annotation steps, and when available, several assistants in the snippet selection and data cleaning process.

The most significant source of uncertainty during data tagging and labelling is attributed to audio prescreening, sound event selection, and annotation as these depend heavily on manual processing, and therefore on human skills and concentration. Specifically, this dependence is conditioned on the degree of familiarity of these experts with the life cycle, behaviour, and diversity of sound emissions of each particular species of interest. Since local dialects and endemic sounds are frequently observed in birds and other species, the overall representativeness of TLs also depends on the amount of field experience in the particular district, where recordings were collected. Therefore, before undertaking the TL creation task, bioacousticians need to acquire personal experience in the area of study and be acquainted with the specifics of local species.

Once the TL creation is initiated (cf. Figure 4.2), the audio prescreening process aims to select an appropriate subset of soundscape recordings for the subsequent processing steps. Here, *appropriate* refers to a purposive selection based on predefined criteria, which are defined according to a certain statistical design of the intended research and technology development activities. For instance, among these criteria could be the requirement for
1. an adequate representation of the seasonal and diurnal patterns of acoustic activity,
2. a balanced representation of all sound categories of interest,
3. an adequate representation of the most significant variability factors of the acoustic environment in the particular habitat, and so on.

The audio prescreening process might not necessary entail scrutinizing each particular recording, and therefore, in general, it is based on random sampling of data and selecting blocks of recordings based on prior knowledge or on certain observational evidence. Hence, the amount of soundscape recordings selected during the audio prescreening process is usually 3–5 times bigger than the desired TL size – the last provides some flexibility in the subsequent data processing steps and facilitates the creation of well-balanced TLs. The overall size of the selected subset of audio depends largely on the quality of recordings, the abundance of the target sound events, and on the requirements of machine learning methods to be used. For instance, generative classification methods and deep learning methods would typically require considerably bigger TLs when compared to discriminative classification methods.

The *snippet selection process* makes use of the subset of soundscape recordings selected during the audio prescreening process. Here, *snippet* refers to a purposely selected portion of audio, which captures one or several target sounds (calls, syllables, songs, etc.) plus a few seconds of the acoustic environment before and after the temporal boundaries of that sound category of interest. These few seconds before and after the target sound provide valuable information about the environmental context. The last is essential for the proper interpretation of sounds during the manual annotation of recordings and is in agreement with the requirements of adaptive noise reduction methods, which often depend on a recurring observation of the acoustic background.

The snippet selection process depends on the availability of confirmed specimen sounds for all categories of interest. Confirmed specimen sounds are provided by scientific data repositories or by experienced bioacousticians and serve as reference signals for the team of assistants that carry out the snippet selection and the hand-editing of audio. The assistants will typically receive support by an experienced bioacoustician or a biologist who supervises the particular activity and helps in resolving difficulties, implements the validation of selected snippets, assumes error correction responsibilities, and so on.

Here, it is important to emphasize that each snippet will possibly contain various sound events of different origin, including sounds of more than one of the categories of interest. Therefore, the compilation of snippets collected at this stage is not appropriate for most of the machine learning methods discussed in Chapter 7 unless audio is cleaned and annotated for the temporal boundaries of each sound event. Some recent advances in multi-label multi-instance classification methods (cf. Section 7.5) offer opportunities for coping with this problem. When the multi-label multi-instance classification methods become computationally scalable and are advanced further in order to cope well with acoustic variability in real-world recordings, the dependence on manual cleaning and annotation of audio will become less pronounced. The last will greatly contribute towards improved scalability of automated species recognition methods and will permit biodiversity monitoring and other related studies to take a great leap forward. However, at present, the mainstream classification methods employed in audio events recognition applications still depend on the hand-editing and annotation steps. Therefore, the workflow shown in Figure 4.2 aims to process the set of snippets in a way, so that these contain only sound events belonging to the category of interest.

The hand-editing of snippets aims to obliterate loud sounds that belong to non-target categories and as much as possible to keep intact only these portions of audio, which contain target sounds. The removal of loud non-target sounds, which overlap with the target categories either in the temporal or frequency domain, is usually performed manually. Most often this is implemented directly on the waveform signal, through some convenient software tool, which allows visualization and editing of the audio spectrogram.[3,4] The main purpose of such editing is to discard all high-energy sound events that do not belong to the category of interest. Afterwards, each hand-edited and filtered snippet can be saved as an individual audio file, which is tagged for the target category of interest. When the noise floor is rather low and when there are no loud competing sounds, such a tagging is sufficient for the needs of machine learning methods and for obtaining reliable statistical models.

In the general case, when TLs are derived from soundscape recordings, the selected hand-cleaned snippets of audio need to be annotated in order to specify the temporal boundaries of all sound events of interest. This is implemented by using a certain audio-editing software tool,[3,4,5] which allows markers and timestamps to be embedded directly into the audio file. Alternatively, one can do the same by listening to the audio (and/or by inspecting the audio spectrogram) and keeping notes about the temporal boundaries of sound events.

The process shown in Figure 4.2 and described earlier is uniform to each target category that has to be modelled.

4.1.4 TLs based on clean recordings

In some cases, the enormous effort associated with the processing of soundscape recordings can be evaded and TLs can be created from clean audio recordings. For instance, when appropriate audio recordings made in laboratory conditions with animals in captivity are available. In that case, each audio recording may well contain sounds of a single species without interference inherent to real-world acoustic environment. Recording animals in captivity is applicable mainly to insects, anurans, and few other groups of species with limited variability of sound emissions and with rigid behavioural patterns.

In general, when all audio recordings are collected in a controlled laboratory set-up, or on the field with a directional microphone, manually pointed out to the individuals of interest, the first three processing steps in Figure 4.2 might not be necessary. In such case, a bioacoustician could straightforwardly segment the audio with respect to the sound emission type and quickly tag each portion of audio. Thus, the tagging of recordings collected in a controlled set-up is much quicker as it does not require the prior processing steps, which are mandatory

---

[3] Audacity, http://www.audacityteam.org/
[4] Praat, http://www.fon.hum.uva.nl/praat/
[5] Adobe Audition, http://www.adobe.com/products/audition.html

when TLs are created from soundscape recordings. Furthermore, manual annotation[6] of sound events is no longer required. If needed, the boundaries of sound evens can be automatically estimated with the help of a simple energy-based acoustic activity detector.

However, in practice it is not feasible to record all species in interest in captivity, so this scenario is not of much help for most biodiversity monitoring and assessment applications. Still it fits some pest control applications and applications concerned with the monitoring of disease-transmitting mosquito species.

4.1.5 Challenges related to TLs' creation

Regardless of whether data collection is implemented in a laboratory set-up or in some other controlled conditions, the manual preparation of TLs is a mandatory part of technology development and an essential step on which depends the advance of bioacoustic methods. However, the required effort and amount of manual labour needed for the preparation of TL does not facilitate the scalability of research and technology development, and thus, it is not well aligned with the aims of computational bioacoustic, where technological support to large-scale biodiversity assessment and monitoring studies is aimed.

Furthermore, as explained before, in the general case when TLs are based on soundscape recordings, the hand-editing and annotation of audio have to be carried out for each species of interest. Consequently, the scale of effort required for the creation of TLs makes it prohibitive in large-scale studies, where identification of many species is targeted.

An unambiguous rule, which would allow us to estimate the amount of effort required for TL creation for each sound category of interest, is not available. Nevertheless, we guesstimate that for recordings with good SNR, the total effort for the manual preparation of a TL requires time investment of at least 10 times the duration of audio data set we need to create. Furthermore, in certain occasions, when extensive hand-cleaning of weak target sounds is required, empirical evidence suggests that the total effort may exceed 20–30 times the duration of audio. All this makes the required effort excessive even when a dozen of sound categories are considered. Furthermore, let us consider the species identification task in large-scale biodiversity studies, where several dozens or hundreds of species are expected. In that case, the effort for TLs' creation would be prohibitive in terms of time and cost for manual labour. The segmentation of soundscapes to snippets

---

6 In the sense of tagging certain audio segments and determining the timestamps of their temporal boundaries.

requires the availability of experts familiar with the acoustic emissions of these species and much more time for the hand-editing and annotations.

4.1.6 Computational bioacoustics in support of TL creation

Given the availability of few sample recordings for each sound category that need to be annotated, computational bioacoustic methods could be used to develop software tools in support of TL creation. Among these are tools that
1. search in audio recordings based on predefined specimen sound event;
2. seek and group together recordings, which meet certain predefined criteria;
3. automatically determine the timestamps of certain predefined category of audio event, and so on.

Furthermore, when certain bootstrap data are available, computational bioacoustics allows the development of category-specific detectors, which could automate the snippet selection process, and thus potentially help in the semi-automatic annotation of long audio recordings.

4.1.7 What if large TLs are not feasible?

When the amount of category-specific audio recordings in a TL is far too small, so that it is insufficient to build a proper statistical model for one or more of the categories of interest, the classification accuracy of the entire identification system deteriorates. One way to cope with this problem is first to create a general acoustic model and then derive the species-specific models from it through maximum likelihood or maximum a posteriori adaptation of the model parameters. The general acoustic model may represent the operational conditions of the environment, in which case it is built from an *acoustic BL*. Alternatively, this general model may represent a neutral sound entity, in which case it can be built from a weighted mix of recordings of multiple categories of sound events (cf. Section 7.3).

4.2 The acoustic background library

When generative statistical pattern recognition methods are considered, most often the general characteristics of the acoustic environment, in which the system will operate, are modelled explicitly with a purposely created acoustic background model (cf. Section 7.2). This acoustic background model, which typically represents the alternative hypothesis H_1, is created based on an *acoustic BL*.

A prerequisite for the creation of a representative acoustic background model is the BL to contain a large amount of soundscape recordings recorded over a long period of time. These recordings have to capture well the acoustic variability at the specific recording environment and to be representative of the acoustic settings of the particular operational set-up. Furthermore, the BL should not contain any audio segments with sound events of the target categories of interest, which will be recognized by the automated system. The last imposes that soundscape recordings need to be manually screened, and in some cases, hand-edited, in order to exclude sound event types that are considered target category in the TLs.

The required size of a BL depends heavily on the daily and seasonal variability of acoustic environment and on the complexity of acoustic conditions in the specific location. Therefore, the minimum required size of a BL may well exceed hundreds of hours of soundscape recordings just in order to capture the most important aspects of acoustic variability.

In Figure 4.3, we present a plausible workflow for the manual creation of a BL. We assume the availability of an audio repository, which contains a large amount of soundscape recordings acquired in the very same operational environment where technology will be deployed.

The creation of a representative BL requires the involvement of an experienced bioacoustician, who is aware of the acoustic variability in the specific habitat and is well acquainted with the sound emissions of the target species. At the first step of BL creation, a bioacoustician will have to select a large number of soundscape recordings that correspond to different acoustic conditions. It is highly recommended that all daily and seasonal variability be suitably represented in the selected subset of recordings. At the next processing step, the selection of soundscape recordings has to be manually screened and hand-edited

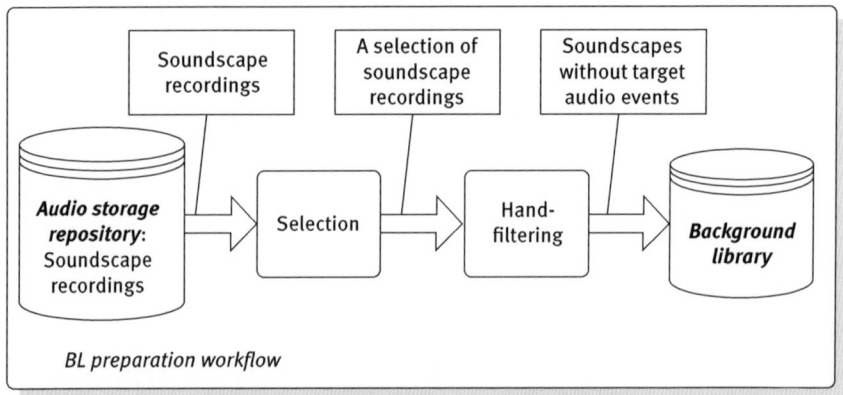

Figure 4.3: Diagram of a plausible workflow for manual creation of a BL.

by a bioacoustician. The last aims to guarantee that sound emissions of all target categories of interest, which have to be recognized and thus are included in the TLs, are eliminated from the BL. If this is not guaranteed, the category-specific model(s) adapted from the general acoustic background model will be with low sensitivity to target sounds, as the log-likelihoods computed with these two models will not differ significantly.

The requirement for manual screening and hand-editing of soundscape recordings demands extensive efforts for prolonged periods. In fact, the manual cleaning of soundscapes is the most time-consuming assignment of the entire BL creation process. Due to the amount of time and effort required, which also translates to high cost due to the involvement of highly qualified experts, sometimes the hand-cleaning of audio is skipped, assuming that the target sound category is not present in the soundscape recordings during certain periods of time. The last might be possible in the case of migratory species or for certain species with acoustic activity in a narrow temporal window. However, in general it is perceived as a suboptimal solution because of the temporal variability of acoustic conditions and the differences that might ensue between these two periods.

Furthermore, when hand-editing of soundscape recordings cannot be avoided, the BL design might be restricted only to the few most critical acoustic conditions, and the rest are ignored in order to reduce cost and efforts.

Having in mind the resources required, the preparation of a well-balanced hand-cleaned BL might be prohibitive, or in many cases unattractive, in terms of expertise needed, effort, or cost. However, the ultimate reward is that a balanced and hand-cleaned BL is a prerequisite for obtaining category-specific recognizers with excellent sensitivity and good robustness under the specific operational environment.

4.2.1 What if large BL is not available?

In some cases, we do not have the raw data, the means, or the resources required for the creation of a general acoustic background model. For instance, when the necessary BL is not readily available and its creation is not feasible within the available project resources or time frame due to lack of soundscape recordings representative of the operational environment, or due to some other reason. In such circumstances, we do not have the option to develop a general acoustic background model. Therefore, there is no way to derive the target category-specific acoustic models via subsequent adaptation based on the background model. In such cases, an alternative could be to consider the cohort model approach, which makes use of a number of competitor models to each target model.

In short, assuming the availability of various acoustic TLs for categories, which are different from the target, we can create a number of cohort models. Each set of cohort models is specific to a certain target model. Typically, one develops a large number of cohort models using all available TLs, and then all cohort models are purposely evaluated for similarity to the target category model, with respect to certain predefined criteria, or a preselected similarity measure. A subset of models that constitute the closest match to the target model is selected for further use. Computational efforts could be reduced when the cohort models are built for just a few carefully selected TLs, representing species or sound events that are known to resemble closely the target category and some that are very different.

During operation of the recognizer, each input chunk of audio is scored against the target and the cohort models. The scores computed for the target model could be normalized with respect to those computed for the cohort models. A final decision is made after applying a threshold on the normalized scores ratio or based on a direct comparison of scores between the target and cohort models.

One disadvantage of the cohort-based approach is that it depends on the availability of multiple TLs, and another is that it might require larger amounts of computations when compared to recognition based on a target model and an acoustic background model. The latter is because a number of cohort models need to be created, based on the corresponding TLs, and then evaluated and ranked. Such process increases the efforts during training when compared to the approach that makes use of a general acoustic background model. However, the computational effort required for training might be comparable or even in favour of the cohort model's approach. Specifically, when we take into account that most often the acoustic background model is based on large amount of data, the lower computational demands might be for the cohort model's creation, evaluation, and ranking. Furthermore, the use of cohort model's approach provides opportunities for a significant reduction of the false-positive error rates as it permits enhancement of the discriminative capability of the recognizer.

The cohort modelling approach is applicable to both the one-category classification tasks as those outlined in Sections 3.2, 3.4, and 3.5 and the multi-category classification tasks such as the species identification, species recognition, and species diarization tasks outlined in Sections 3.3, 3.6, and 3.8.

4.2.2 Computational bioacoustics in support of BL creation

As discussed in the previous subsection, a category-specific recognizer (cf. Section 3.4) does not necessarily depend on the availability of a BL, that is, the recognizer could be implemented based on the cohort modelling approach. Once such a category-specific recognizer is available for the target categories of interest, it could be used for a significant speed up of the BL creation. This speed up would result mainly due to the reduced effort for selective hand-editing of soundscape recordings (cf. Figure 4.3), meaning that the discovery of end elimination of sound events of the target species can be implemented automatically or semi-automatically with the help of the species-specific recognizer.

Therefore, an automated or semi-automated screening and selection of audio recordings (such functionality was outlined in Section 3.10) would greatly facilitate and speed up the creation of an acoustic BL. Such an automation depends on the availability of rule-based tools or search-by-example tools, which can discover the target species sound events and eliminate them from the soundscape recordings, or alternatively only select files which do not contain sound events of the target species. The subset of recordings that remain after such a selection and/or editing are assumed clean of the target species, and thus can be used as a BL for the creation of an acoustic background model. The benefits in favour of the efforts to create and use an acoustic background model are quite significant. The acoustic background model can

1. model the alternative hypothesis;
2. help in the creation of a representative target model, even when the TL is small; and
3. create representative model of everything that is not the target category, that is, helps in implementing an exhaustive modelling of all possibilities.

Semi-automated editing of soundscape recordings would mean that an automated recognizer tags (and timestamps the boundaries) of all sound events that need to be eliminated from the BL. Next, a bioacoustician will use a graphic user interface to inspect visually the audio spectrograms for all groups of tagged sound events, or when needed all recognized instances marked for deletion, and will eventually make a decision whether these chunks will be kept or will be eliminated from the BL.

Following akin workflow (cf. Figure 4.3), the semi-automated approach possesses the potential to decrease considerably the efforts needed for the creation of a BL. As already said, in such cases the BL creation process is reduced to preselecting a subset of soundscape recordings for further processing by the recognizer and then an expert makes an instance-by-instance or group-by-group decision as to which sound events are to be eliminated and which are to be kept. For the reason that a human expert is involved only in the inspection of certain automatically selected portions of audio, the semi-automated approach would entail a lesser demand of time and effort when compared to the case when listening and manual editing of the entire selection of soundscape recordings is required.

Therefore, instead of listening to and screening the spectrograms of hundreds of hours of soundscape recordings, which is the typical range for the size of a BL, a human expert will listen to a few per cent of the audio, and this is only when all instances need to be manually checked. In the case when the decision can be made on a group-by-group basis, the human expert only needs to listen to one example of each group (usually the one which is closest to the geometric mean of all group members) or just a few randomly selected examples in order to get an impression about the within variability of each group.

4.3 The validation library[7]

Once all category-specific models are built (making use of certain TLs and sometimes of a BL), a VL is required in order to fine-tune the adjustable parameters of the recognizer and optimize the overall performance. Such a fine-tuning is very important as it adapts the settings of the audio processing tool to the particular acoustic conditions in which technology will operate. The type of fine-tuning depends on the machine learning method used and the nature of the particular task. For instance, for the tasks described in Sections 3.2–3.10 and other related tasks among the adjustable parameters could be

---

[7] The *validation library* is used for the purpose of technology development, and therefore, its function largely corresponds to the one of a *development data set* (conforming to terminology used in the machine learning community). However, here we will distinguish between *validation library* and *development data set* because these may have different origin and different build. In particular, with the designation VL we unambiguously define both purpose and content of the data set, and this will be our main motivation to prefer it to the more broad and underspecified term *development data set*. The term *development data set* describes the purpose and use of data; however, it does not define strictly (or disclose) its build and content. In brief, both the *validation library* and the *development data set* are used to estimate the adjustable parameters of a system or to train a hierarchical multi-stage classification scheme. In particular, the development data set is sometimes derived by keeping aside a portion of the training data set. However, this is not a valid option when concerns the validation library. The problem here is that a VL should contain only *raw unprocessed soundscape recordings*, which are annotated for the sound events of interest. In contrast, training libraries typically consist of short snippets, which are hand-edited, filtered, and usually only tagged for the target category of sound events. The last means that (i) usually TLs are heavily processed, (ii) might not be annotated, and (iii) might lack timestamps for the sound events of interest. In contrast, a VL would always contain time-stamped sound events annotated directly in raw unprocessed soundscape recordings. Therefore, a VL cannot be derived as a subset of a TL as the development data sets sometimes are.

1. any kind of decision threshold,
2. the parameters related to the sensitivity and selectivity of the recognizer,
3. the speed of adaptation to changes in the noise floor levels,
4. the adjustable parameters of the noise suppression stage, and so on.

Obviously, the VL should be representative of the acoustic environment in which technology will be deployed to operate. This is a prerequisite for achieving a high recognition accuracy, as the primary purpose of a VL is to serve as reference for fine-tuning the adjustable parameters of an audio processing tool. This is only possible when the VL is based on soundscape recordings collected in the particular acoustic environment. Therefore, we should better think of the *VL* as a purposely selected subset of duly annotated soundscape recordings, which are representative of the (i) acoustic conditions in the operational environment where the audio recording happened and (ii) of the sound events which the recognition tool will process.

The major difficulty when creating VLs comes from the requirement of the selected subset of soundscape recordings to be annotated for the temporal boundaries of the target sound events.[8] The last is required so that we can estimate the recognition accuracy of the audio recognizer for each setting of the adjustable parameters.

In Figure 4.4, we show a workflow for a VL creation. In brief, assuming the availability of an audio repository with a large number of raw unprocessed soundscape recordings, we can build a VL by following a two-stage process: (i) the selection of a representative and well-balanced subset of soundscape recordings and (ii) their annotation. The quality of VL will largely depend on the skills of an experienced bioacousticians, who is involved in these processing steps.

---

8 In contrast, in the audio event recognition evaluation campaigns (Giannoulisy et al. 2013; Stowell et al. 2015; IEEE AASP Challenge 2016), the *development sets* might consist of synthetic sequences of events. In this way, temporal boundaries of events are known and the ground true annotation is readily available. For instance, in the speaker recognition evaluation campaigns (NIST SRE 1996–2012), development of data sets typically consists of tagged recordings, and no annotations are needed. In the two-speaker recognition task, the turns of speakers are known in advance before mixing the two audio channels, so annotation is avoided. However, in the speech/speaker diarization of meetings (NIST RT 2002–2009), annotations are mandatory as they provide the ground true reference against which technology is evaluated and optimized. All the above-mentioned evaluation campaigns consider different application set-ups, and therefore use different builds of the development data set. However, a validation library would always be annotated and include the timestamps for the acoustic events of interest.

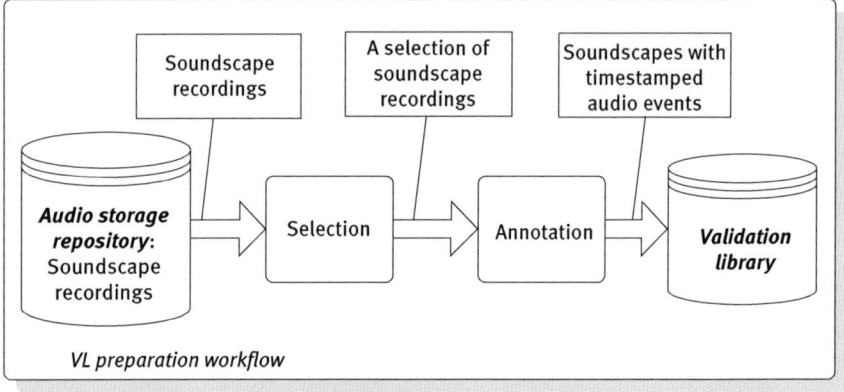

Figure 4.4: Diagram of a plausible workflow for manual creation of a VL.

In particular, a bioacoustician will first need to design the statistical plot of the VL and then select a balanced subset of soundscape recordings. Each recording in this subset will contain sound events representative of both the target category/categories of interest and the main sound sources responsible for the bulk of false-positive errors. Next, the selected subset of soundscape recordings is annotated so that all sound events of interest are tagged and their temporal boundaries are timestamped. The tags and timestamps obtained to this end serve as the ground true reference against which the overall recognition performance will be assessed. The last is important because the recognizer fine-tuning is usually performed in an iterative manner. This implies that the audio processing tool is repeatedly evaluated, each time with different settings of the adjustable parameters, until an acceptable performance is reached.

The VL should be of a limited size but still representative of the acoustic conditions in the operational environment. A representative VL would provide the basis for approaching the optimal settings during the iterative fine-tuning of the adjustable parameters. A large VL would provide a more accurate estimation of performance; however, the overall effort for VL creation and for tuning of the audio tool might become overwhelming or even prohibitive when multi-class identification set-up is considered.

In this regard, the typical size of a *VL* is in the range of an hour to few hours of audio. This means that the VL is always much smaller than a BL. However, in contrast to a BL, a VL does not require hand-editing and filtering of audio because it has to be kept raw, pristine, and representative of the real-world acoustic conditions. Therefore, the main challenge related to the creation of VL is that it has to be properly annotated for the sound events of the target category.

These annotations (both tags and timestamps) are assumed the *ground true* reference against which the recognition results are compared in order to estimate the recognition performance.

The requirement for unaltered annotated soundscape recordings makes the VL much different from TLs (Section 4.1) and BL (Section 4.2), and therefore it cannot be considered akin or derivative to neither of these.

4.3.1 Computational bioacoustics in support of VL creation

Due to the specific purpose and design requirements of VLs, the VL creation workflow does not favour full automation. A good VL design involves achieving statistical balance among the requirements for
1. a small overall size,
2. a good representation of the main acoustic variability factors of the operational environment,
3. a good representation of the target sound categories,
4. a good representation of the major sources of false-positive (false alarm) errors, and so on.

Furthermore, once the statistical design is ready, an experienced bioacoustician has to listen and manually screen the audio spectrograms of a large number of soundscape recordings in order to select an appropriate subset, which conforms to the statistical design.

Nevertheless, certain technological support for semi-automated processing of audio might reduce the overall effort for VL annotation. For instance, when a recognizer for the target sound category is already available, it can be used to process the selection of soundscape recordings in order to tag the sound events of interest and provide an automated estimation of their timestamps. The last will not liberate the human expert from listening again to the entire subset of soundscape recordings; however, it might reduce the effort for manual introduction of temporal boundaries from scratch.

For instance, when the automated acoustic recognizer succeeded with both the correct tags and an accurate estimation of the time boundaries of a certain sound event, the bioacoustician would simply acknowledge this and promptly move to the next sound event. In the cases when a certain tag is correct but its timestamps are not, or the opposite, when the tag is not correct but its timestamps are, the effort for correcting these errors is assumed smaller than the initial effort for data annotation from scratch. Unfortunately, when both a tag and its timestamps are not correct, the automated processing will introduce an extra burden and will contribute to distraction of the human expert. Therefore, the whole effort for semi-automated annotation of soundscape recordings makes sense only when the acoustic recognizer used for that purpose has a relatively good recognition accuracy, and when the soundscape recordings are not that challenging, so the amount of errors remains small.

4.4 The evaluation library

Once the performance of a certain recognizer or another audio processing tool is optimized, and the adjustable parameters fine-tuned to the operational environment, it is ready for deployment and exploitation. Before that, however, we need to obtain a fair assessment of the system performance in conditions matching the real-world operational environment. The EL provides the means for such an assessment. For that purpose, the EL is designed as a selection of annotated soundscape recordings which are collected in the very same acoustic environment at the location where the system will be deployed, or when this is not feasible, in an equivalent acoustic environment. In order to ensure a fair evaluation of the recognition performance, the EL has to match the operational set-up and closely approximate the operational conditions.

The workflow used for the creation of an EL is identical to the one of VL (cf. Figure 4.4); however, the two acoustic libraries have to be independent and selected without overlapping among the soundscape recordings. This is because any overlapping between the soundscape recordings used for the creation of EL and VL would result in an overoptimistic assessment of system performance. The last will increase the risks of incorrect interpretation of recognition results and of misunderstanding the underlying processes in the monitored areas.

As shown in Figure 4.4, the process of EL creation assumes the availability of a large repository of soundscape recordings captured in the operational environment. At the first processing stage, a bioacoustician selects a subset of soundscape recordings that represent well the acoustic environment in which the audio processing tool will operate. This selection of soundscape recordings becomes subject to a careful annotation, which aims to identify and tag all sound events belonging to the target categories of sounds (these present in the TLs) and their temporal boundaries. Both tags and their timestamps are considered the ground true reference against which recognition results will be compared in order to measure the overall recognition performance. The error rates computed based on the EL provide an important information about the expected behaviour of automated audio processing tool in real-world conditions. The evaluation results obtained through the EL provide insights about certain weakness and biases of the automated recognizer. Therefore, one might be interested to analyse in depth the cases where errors appear and evaluate any bias introduced by the recognizer with respect to the ground true annotation. A proper understanding of the weaknesses and biases introduced by technological tools will help in the planning for subsequent improvements of methods and technology.

More importantly however, the evaluation results reveal how the acoustic tool will perform in real-world operational conditions, which is quite necessary for a proper interpretation of results during practical use in any application scenario involving biodiversity monitoring or biodiversity assessment. Such an understanding is indispensable because no automated system is perfect in terms of recognition accuracy and precision. Furthermore, the overall recognition performance would usually vary depending on the fluctuation of numerous factors in the operational environment. Therefore, for the proper interpretation of empirical results it is quite important to have a clear understanding about the specific limitations and biases introduced by the particular audio recognition tool. Obtaining a fair estimation of precision and accuracy in real-world conditions, preferably in the specific application scenario and at the location where it will be deployed for operation, is only the lowest level of such an understanding. A higher level of understanding would mean studying the dynamics of errors in different conditions and devising a procedure for mitigation of the bias introduced by the recognition errors.

The EL usually has a much larger size when compared to the VL, and both the VL and the EL remain much smaller than the BL. The main difference between the VL and the EL is in their purpose. An EL is not used in the system development but only for realistic evaluation of recognition performance. This implies that the EL is used only once, and therefore its larger size will not create significant computational burden as it could be in the case with the VL. Furthermore, a larger EL allows for achieving a better resolution of the system performance assessment, and therefore a larger size of EL, is always welcome. This is the primary reason why EL is preferably made several times bigger than VL. The actual size of EL depends on the particular application task (including those outlined in Sections 3.2–3.10 and other sections) and on the complexity and number of sound categories that need to be recognized. The size of an EL is usually in the range of few hours of carefully selected and annotated soundscape recordings.

4.4.1 Computational bioacoustics in support of EL creation

As with the VL, the main difficulty in the creation of EL is the need of duly annotation of all sound events belonging to the target categories of sounds. Similarly to the VL, a semi-automated approach could be used to speed up the EL annotation. The last is conditioned on the availability of operational recognizer for the target sound categories of interest. Despite the automated processing, a human expert will still listen to the soundscape recordings for the entire EL and will manually inspect their audio spectrograms in order to correct errors in the tagging and determination of timestamps. Any flaws in the EL implementation would translate into a bias in the overall performance assessment and might lead to misinterpretation of the underlying processes in the monitored habitat.

Concluding remarks

As highlighted throughout this chapter, technological support with intelligent automated or semi-automated data annotation tools is highly desirable for the development of all kinds of acoustic libraries: *TL*, *acoustic BL*, *VL*, and *EL*. However, given that these libraries do not exist and have to be created from scratch, the development of such resources also depends on a certain amount of carefully selected and annotated audio recordings, referred to as *bootstrap data set*. In fact, the bootstrap data set consists of a very limited amount of manually annotated recordings, a subset of a larger acoustic library, which is under development. Usually, the bootstrap data set is used for the creation of an initial (provisional) model, which is afterwards enhanced iteratively until certain performance goals are satisfied or until some stop criterion is met. Typical stop criteria are
1. lack of progress after certain a number of iterations,
2. a predefined total number of iterations is reached,
3. the objective error criterion is satisfied, and so on.

Once this provisional model reaches a satisfactory performance, it can be employed in an acoustic recognition tool that will perform semi-automated processing of soundscape recordings. In some cases, the technological support provided by such a simple tool could bring quite a significant reduction of the overall time and efforts required for the development of acoustic libraries. This is mainly due to the reduced time for preselection of audio files and for providing automated annotations.

Unfortunately, at the present stage of technology development, human experts still play a central role in the preparation of ground true annotations for the sound events of interest (tags and their temporal boundaries), which impedes proper technological support and scalability of biodiversity assessment and monitoring.

5 One-dimensional audio parameterization

Introduction

The 1-D audio parameterization methods are well established and widely used in many of the computational bioacoustic tasks outlined in Chapter 3, and more specifically, those related to one-species detection (Section 3.1), species identification (Section 3.2), one-category recognition (Section 3.4), one-species recognition (Section 3.5), sound event-type recognition (Section 3.8), and so on. However, the use of 1-D audio parameterization methods in multi-label species identification task (cf. Sections 3.3) is somehow complicated. The problem is that the 1-D methods typically compute audio descriptors that enclose information about all frequency bands within a certain predefined range of the spectrum, and there is no simple way to resolve overlapping of different sound events in the frequency domain. In the following, we explain the sequence of processing steps common to many 1-D audio parameterization methods.

In short, in the 1-D audio parameterization, the signal is split into a sequence of short segments, each assumed to represent a stationary portion of the audio. For each segment, a set of audio descriptors is computed so that the relevant information is represented with a low-dimensional feature vector, and the irrelevant information is suppressed. The feature vectors computed for a sequence of audio segments are normally post-processed before they are fed to the pattern recognition stage. In summary, the audio parameterization process aims to
1. compensate for undesired variability in the signal;
2. suppress certain interferences;
3. reduce the number of descriptors; and
4. compute derivative descriptors, which carry out information about the temporal evolution of the original raw descriptors, and therefore, complement the static feature vector with dynamic information.

The main audio parameterization stages include *audio signal pre-processing*, *audio descriptors computation*, and *post-processing of the audio descriptors* (Figure 5.1). In brief, the audio *pre-processing* stage aims to reduce the negative effects due to environmental noise and when possible lessen the influence of loud interferences from co-occurring sound events. The well-conceived audio pre-processing is essential to the computation of informative and reliable audio descriptors. Subsequently, the *post-processing* of audio descriptors aims to suppress the major sources of undesirable variability, change the feature vector length, add dynamic information, and so on. Common sources of

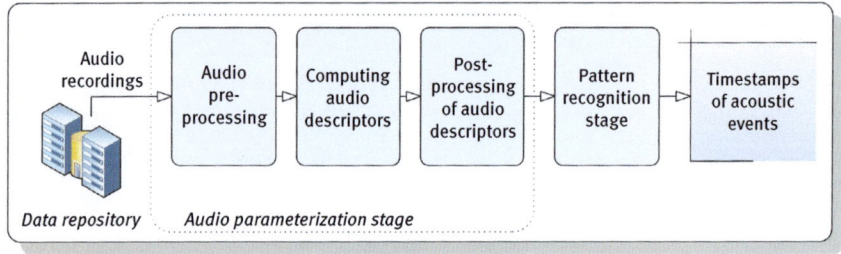

Figure 5.1: An overall block diagram of the audio parameterization process.

variability are linked to seasonal or weather-related changes in the environmental conditions, different equipment set-up, time-varying (floating) noise floor level due to changes in the acoustic activity of other species, and so on.

In the following sections, we provide details of the signal processing stages. It is common for these to have different implementations and varieties; however, the overall concept is characteristic of a wide range of discrete Fourier transform (DFT)-based cepstral coefficients. Among these are the traditional Mel-scale frequency cepstral coefficients (MFCCs), linear frequency cepstral coefficients (LFCCs), Greenwood function cepstral coefficients (GFCC), and other akin signal parameterization schemes. Here, we will only discuss these in general and will not get into details on the variations due to different warping of the frequency axis, dissimilar filter bank design, magnitude transformation law, and so on. The interested reader might refer to Ganchev (2011) for a comprehensive description of 11 DFT- or wavelet packet transform-based audio parameterization methods and their empirical evaluation on three speech-processing tasks.

5.1 Audio pre-processing

The audio signal pre-processing stage typically consists of *pre-filtering*, *segmentation*, and *windowing* steps (Figure 5.2). The *pre-filtering* of the input signal, $s(n)$, most often consists of mean value removal, bandpass filtering, and, when necessary, downsampling and noise suppression. The mean value removal aims at eliminating any signal offset, which might be due to the audio acquisition device or the subsequent data-coding algorithm. The bandpass filtering is often seen as a simple noise reduction mechanism which is used to improve the signal-to-noise ratio (SNR) before the actual audio parameterization. It typically preserves the frequency range where most of the target species signal energy is located and reduces the influence of additive interferences from the environment. For instance, interferences due to wind blowing in the microphone, mechanical vibrations provoked by wind or other source, vehicle sounds due to human

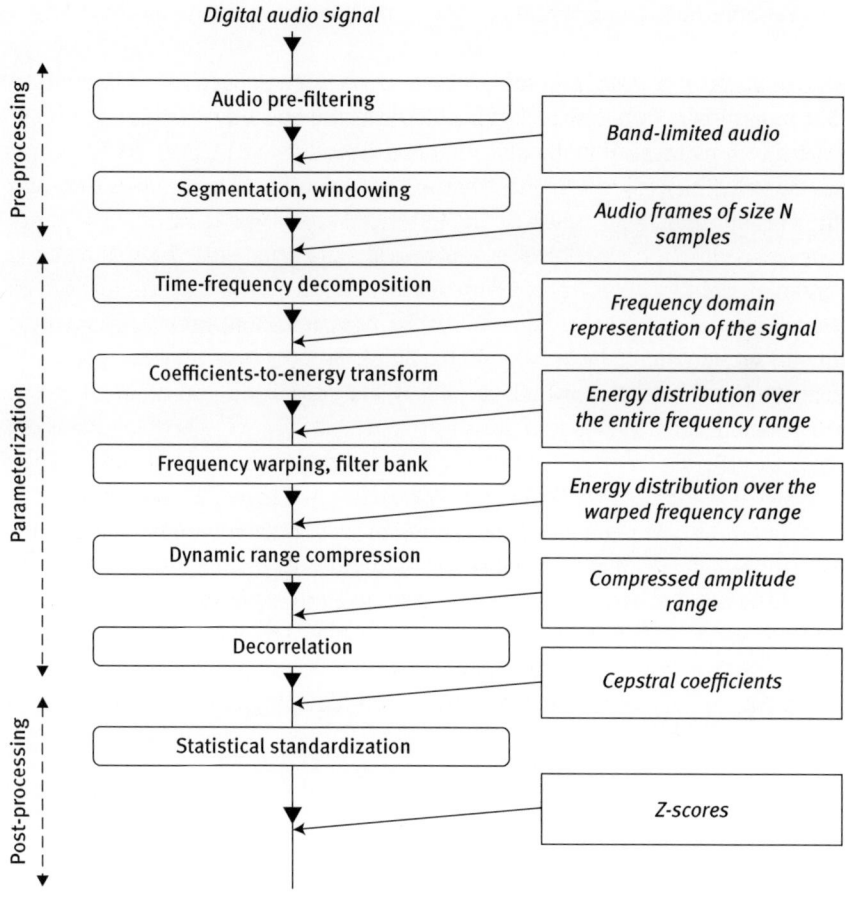

Figure 5.2: Block diagram of the audio parameterization process.

activity, low-frequency sound emissions from other animals, and so on. Except for the attenuation of the direct current (DC) offset and low-frequency interferences, the bandpass filtering also reduces the contribution of the high-frequency components that might occur in clipped signals. Where feasible, the downsampling process is used to decrease the computational demands during the subsequent processing steps.

The *signal segmentation* step builds on the assumption that the spectral characteristics of the audio signal do not change significantly for the duration of a short segment. The signal segmentation process typically results in overlapping segments with a fixed size. However, in many cases the use of non-overlapping variable-length segments provides a convenient way for preserving the integrity of sound events and reducing the overall computational demand. For instance, this is the case with many insect and frog sound emissions.

5.1.1 Variable-length segmentation

The variable-length segmentation is based on a detector of acoustic activity that is fed by an energy estimator, which computes the short-term energy E_{ste} for a segment of K successive samples as

$$E_{ste}(p) = \sum_{i=1}^{K} \left(\breve{s}(pL + i) \right)^2, \quad p = 0, 1, \ldots, P - 1. \tag{5.1}$$

Here, $\breve{s}(.)$ is the pre-processed input signal, p is the group index, L is a predefined step size which defines the degree of overlapping between two successive segments, and

$$P = \text{fix} \frac{(N - K + L)}{L} \tag{5.2}$$

is the number of segments in a recording with length N samples. The operator fix stands for keeping the integer part of the real number, that is, represents rounding towards the smaller integer number. The $E_{ste}(p)$ contour is further used as input to the detector of acoustic activity

$$D(p) = \begin{cases} 1 \text{ OR } D(p-1) & \text{for } E_{ste}(p) \geq \theta_{high} \\ 0 \text{ AND } D(p-1) & \text{for } E_{ste}(p) < \theta_{low} \end{cases}, \tag{5.3}$$

where OR and AND are Boolean algebra operators, $D(p-1)$ is the value of $D(p)$ for the preceding group, and the thresholds $\theta_{high} = k_{high} E_{bkg}$ and $\theta_{low} = k_{low} E_{bkg}$ control the sensitivity of the acoustic activity detector for the start and end points, respectively. For the sensitivity control constants k_{low} and k_{high}, it typically holds that $k_{high} \geq k_{low}$. Finally, E_{bkg} stands for the estimation of the averaged mean energy of the signal for segments labelled as background acoustic activity.

In brief, the variable-length segmentation algorithm operates as follows: In its relaxed state, the activity detector $D(p)$ is set to zero, that is, no acoustic activity is detected. When the level of acoustic activity increases, which causes $E_{ste}(p)$ to reach or to exceed the threshold θ_{high}, the activity detector $D(p)$ triggers to value 1. It remains in that state until the value of $E_{ste}(p)$ drops below the threshold θ_{low}, when the activity detector relaxes to its inactive state, $D(p)$ is set to 0.

Since the subsequent estimates of energy are for overlapping groups, the precision of detected onset and end borders depends on the step size L. Thus, L is selected to provide a reasonable trade-off between temporal resolution and computational demands. For obtaining a smooth estimation of the short-term energy E_{ste}, the group size of K samples is typically selected in the range of milliseconds.

5.1.2 Uniform-length segmentation

In the process of segmentation with uniform segment size, a window function $W(.)$ that is shifted with a certain step L is multiplied with the band-limited audio signal (cf. eq. (5.4)). The window function is defined with a limited number N of non-zero elements, depending on the desired segment size (eq. (5.5)). The window function could be of any shape (Harris 1978); however, most often the Hamming window is used. This is because it provides a reasonable trade-off between effective window size in time-domain and frequency-domain resolution, in terms of bandwidth of the main lobe, suppression of the first side lobe, side lobe fall-off.

The product of the long audio signal $N_{sig} \gg N$ and the window function of length N, shifted on lL samples, results in the formation of l uniform audio segments, each with length of N samples. Due to practical reasons, neighbouring audio segments are usually overlapped by 50 % or 75 %, that is, the skip step size of audio parameterization is $L = \text{fix}(0.5N)$ or $L = \text{fix}(0.25N)$ samples. However, in general the value of L is not coupled to the window length N, and thus L could take any integer value, including those for which the neighbouring segments do not overlap at all. Still for the skip step size L it holds true that lesser is better. This is because L, together with the sampling frequency F_s, defines the lower bound of temporal resolution for timestamps. Smaller values of L provide the opportunity for a more accurate estimation of begin and end timestamps of detected sound events.

5.2 Audio parameterization

Next, the audio signal, which is pre-processed and segmented to uniform- or variable-length segments (cf. Section 5.1), is subject to audio parameterization process (Figure 5.2). In the following, we outline the processing steps that are typical of the traditional MFCCs, the LFCCs, and other DFT-based methods for the computation of cepstral coefficients.

Each audio segment $\breve{s}(n)$, $n = 0, 1, \ldots, N - 1$, consisting of N samples of the pre-processed signal, is subject to the short-time discrete Fourier transform (STDFT):

$$S_l(k) = \sum_{i=0}^{N-1} \breve{s}(i + lL) W(i) \exp\left(-\frac{j2\pi i k}{N}\right), \quad 0 \le i, k \le N - 1, \tag{5.4}$$

where i is the index of the time domain samples, k is the index of the Fourier coefficients, l is the index of the current audio segment, and L denotes the displacement (skip step size, aka skip rate) of the current segment in terms of number of samples with respect to the previous segment. The actual choice of L corresponds to selecting a particular trade-off between required computational demands and desired temporal resolution, and thus depends on the specific application scenario. In the case of Hamming window, the window function $W(.)$ is defined as

$$W(m) = 0.54 - 0.46\cos\left(\frac{2\pi m}{N}\right), \quad m = 0, 1, \ldots, N-1. \tag{5.5}$$

It is applied to reduce the spectral distortions caused by an abrupt change of signal amplitude at the boundary points of the audio segment.

Once the Fourier coefficients are computed (eq. (5.4)), we can implement some nonlinear warping of the frequency axis and a nonlinear compression of the magnitude $|S_l(k)|$ or the power spectrum $|S_l(k)|^2$. These nonlinear transformations aim to approximate the nonlinear perception of loudness and frequency in mammal auditory system. The frequency warping is implemented through a filter bank, and the nonlinear compression of magnitude in mammal auditory system is approximated by logarithmic, cubic, or another nonlinear law. Since the following steps apply for every segment, in the rest of this chapter, we omit the subscript l but it remains implied.

The use of a filter bank aims to reduce the frequency resolution of spectral representation, so that we can obtain a more compact and less noisy set of audio descriptors at the subsequent audio parameterization steps. Depending on the particular design, the filter bank might be linear or may warp the frequency axis according to some predefined nonlinear function, such as the Mel-scale or another species-specific frequency warping scale (Greenwood 1961, 1990, 1996, 1997). The choice of frequency warping function is an application-specific issue.

The selection of the frequency range of interest is usually implemented by adjusting the lower and upper boundaries of the filter bank. Depending on the particular application, the filter bank may cover only the species-specific frequency range, some wider bandwidth of interest, or the entire available bandwidth of the signal. However, most often, the frequency range of the filter bank is selected to encompass the predominant portion of signal energy, computed for the sound events of interest, and exclude the strongest sources of interference when this is feasible.

The shape of the individual filters in the filter bank is most often selected triangular for the sake of simplicity; however, there are implementations that use filter shaping, which approximates the auditory filters of mammal auditory system.

In the case of linear spacing filter bank $H_i(k)$, which consists of B equal bandwidth and equal height triangular filters, we have

$$H_i(k) = \begin{cases} 0 & \text{for } k < f_{b_{i-1}} \\ \frac{(k - f_{b_{i-1}})}{(f_{b_i} - f_{b_{i-1}})} & \text{for } f_{b_{i-1}} \le k \le f_{b_i} \\ \frac{(f_{b_{i+1}} - k)}{(f_{b_{i+1}} - f_{b_i})} & \text{for } f_{b_i} \le k \le f_{b_{i+1}} \\ 0 & \text{for } k > f_{b_{i+1}} \end{cases}, \quad (5.6)$$

where $i = 1, 2, \ldots, B$ stands for the ith filter, f_{b_i} are the boundary points of the filters, and $k = 1, 2, \ldots, N$ corresponds to the kth coefficient of the N-point DFT. The boundary points f_{b_i} are expressed in terms of position, which depends on the sampling frequency F_s and the number of points N in the DFT.

Here, the centre of the first filter is defined at the lowest frequency in the spectrum of sound emissions for the particular target species. The centres of subsequent filters are displaced linearly with a predefined step and define the boundaries of their neighbour filters. Specifically, each filter starts from the centre of its left neighbour and ends in the centre of its right neighbour filter.

In other implementations of the filter bank, the filter bandwidth is decoupled from the centres of the neighbouring filters (Skowronsky and Harris 2004), which allows the lower frequency filters to become narrower and the high frequency wider than in the other filter bank designs. This was reported to enhance noise robustness in low SNR conditions (Skowronsky and Harris 2004) and improve recognition accuracy when distinguishing between individuals (Ganchev et al. 2005). Both suggest that this filter bank design and the corresponding human-factor cepstral coefficients (HFCC) might be advantageous also to the one-species detection, the one-species recognition, species identification, and other related tasks (cf. Chapter 3).

Usually, the linear spacing (equal frequency resolution among all frequencies) is a reasonable choice because many animal species produce wide-band sounds which are spread over the entire frequency range. However, nonlinear spacing between the filter centres, and nonuniform width of filters, motivated by the hearing apparatus of animals and humans is frequently used. The majority of studies currently rely on the use of MFCCs, which are considered appropriate for audio processing applications, including the tasks related to acoustic recognition of species (cf. Chapter 3). A comparison between various implementations of the MFCC and other audio parameterization techniques, which employ nonlinear filter spacing and non-uniform filter bandwidth, is available in Ganchev (2005, 2011).

Instead of the Mel-scale-based warping of the frequency axis, the Greenwood function (Greenwood 1961, 1990, 1996, 1997) has been used too. It offers a generalized representation of the frequency warping according to the hearing capabilities of various animal species. Audio features based on the Greenwood function, referred to as GFCC (Clemins and Johnson 2006; Clemins et al. 2006), have been used in species-specific studies on automated acoustic recognition (Steen et al. 2012).

Regardless of specific type of frequency warping implemented by means of the filter bank, in the next processing step we perform decorrelation of the log-energy filter bank outputs S_i, where

$$S_i = \log_{10}\left(\sum_{k=0}^{N-1} |S(k)|^2 \cdot H_i(k)\right), \quad i = 1, 2, \ldots, B, \tag{5.7}$$

via the discrete cosine transform (DCT):

$$x_j = \sum_{i=1}^{B} S_i \cos\left(j(i - 0.5)\frac{\pi}{B}\right), \quad j = 0, 1, \ldots, D. \tag{5.8}$$

Here, j is the index of the LFCC, and D is the total number of LFCCs that are computed, where $D \leq B$.

Since the above analysis applies to every audio segment, it is obvious that the output of the audio parameterization stage is a sequence of LFCC vectors. We use $\mathbf{X}_{all}$ to represent a sequence of LFCC vectors computed for a sequence of audio segments of the original audio recording. Therefore, in the general case, $\mathbf{X}_{all}$ is a sequence of T D-dimensional audio feature vectors, $\mathbf{X}_{all} = \{\mathbf{x}_l\}\big|_{l=1}^{T}$, where $\mathbf{x}_l = \{x_j\}\big|_l$ is the vector computed at a certain time instance, equal to the product of the sampling period, the window skip rate L, and the index l, and T is the total number of audio feature vectors.

5.3 Post-processing of audio features

The audio features *post-processing* can be categorized in two broad categories aiming at (i) eliminating some undesired variability of the audio features, excluding non-informative features and less-promising segments, or (ii) implementing some feature space transformations, which would facilitate the creation of robust and more sensitive statistical models at the subsequent machine learning stage. In the first category fall post-processing methods, which aim to
1. reduce the undesired influence of time-varying acoustic environment conditions,

2. derive a more compact representation of the information encoded in the audio feature vector,
3. exclude from further processing the non-promising regions of audio signal, and so on.

The post-processing of audio features, such as cepstral liftering, statistical standardization of all values, decorrelation of audio descriptors, excluding non-relevant audio features, and so on, belongs to the second category of post-processing methods, which aims at facilitating the training process or obtaining a better-quality statistical models.

In general, post-processing may include but is not limited to
1. dynamic range standardization for the individual audio features;
2. a (non)linear modification of the magnitude and/or the statistical distribution of audio features, in order to filter out certain undesired and/or non-informative variability, or to change the statistical properties of audio features;
3. audio feature ranking and selection of the most informative audio features for a specific application;
4. subspace mapping or another nonlinear transformation, which aims to decrease the number of audio features, that is, transformation to another set of features that has a certain advantage, such as a more compact representation, more robust to rounding errors, which allows decoupling of information sources, and so on;
5. temporal smoothing (aka temporal filtering), which limits the speed of magnitude change during temporal evolution of descriptors;
6. computing temporal derivatives of the audio features in order to reveal short-term and/or long-term temporal correlations in their sequence;
7. filling in missing values;
8. eliminating audio feature values computed from unreliable audio segments;
9. generating statistical descriptors for an audio chunk or for an entire recording in order to uncover hidden properties of the audio descriptors computed on a segment level; and
10. averaging audio descriptors over time for entire syllable or for an audio chunk for obtaining a compact, length-independent representation of a sound event.

In the following sections, we focus on the statistical standardization of audio features and on the computation of temporal derivatives as these have a well-documented record of benefits in practical applications and facilitate the training of statistical models in a wide range of pattern recognition algorithms.

5.3.1 Statistical standardization

Statistical standardization of audio features means that all parameters in a sequence of feature vectors are to be normalized to zero mean value and unit standard deviation. This results in obtaining the z-score $\bar{x}_j$ of each individual parameter x_j of the audio feature vector:

$$\bar{x}_j = \frac{x_j - v_j}{\xi_j}, j = 0, 1, \ldots, D. \tag{5.9}$$

The z-scores $\bar{x}_j$, computed based on the audio feature vectors, quantify the distance from the mean value in terms of the standard deviation. The mean value v_j and the standard deviation ξ_j are estimated on per recording basis for each of the D dimensions, that is, for each audio descriptor.

5.3.2 Temporal derivatives

The temporal derivatives allow extracting some information about the dynamics in a sequence of feature vectors. The first temporal derivative (aka delta coefficients) indicates the speed of change in the amplitude of audio descriptors. It could be computed through the simple difference:

$$\Delta \bar{x}_{j,l} = \frac{\bar{x}_{j,l+\tau} - \bar{x}_{j,l-\tau}}{2\tau}, j = 0, 1, \ldots, D, \quad l = 1, 2, \ldots, T, \tau = 1, 2, \ldots, Q, \tag{5.10}$$

where τ is an integer specifying the time shift in terms of number of segments with respect to the current segment index l, and $Q = 2\tau + 1$ is usually a small number specifying the largest desirable time span in terms of feature vector indices l. The bigger the time shift τ, the more distant temporal relations are sought. In practice, when computing short-term temporal relations, the values of τ are in the range between 1 and 5 depending on the degree of overlap between neighbouring audio segments and the desired temporal span.

For instance, if $\bar{x}_{j,l}$ is computed for an audio segment at time l, we estimate the first derivative $\Delta \bar{x}_{j,l}$ by using its previous $\bar{x}_{j,l-1}$ and next $\bar{x}_{j,l+1}$ temporal neighbours. The boundary value for $l = 1$, that is, the first value in a sequence of feature vectors, is set by duplicating the value computed for $l = 2$, and the value for the last for $l = D$ is set to the one computed for $l = D - 1$.

However, the first temporal derivative computed through the simple difference (5.10) is known to be noisy due to random fluctuations of audio descriptors computed for subsequent segments. Therefore, in practice we make use of a smoothed version of the first difference, which involves temporal smoothing of the trajectories of the individual audio features. Most often, this is implemented with some regression smoothing function, such as

$$\Delta \bar{x}_{j,l} = \frac{\sum_{\tau=1}^{Q} \tau (\bar{x}_{j,l+\tau} - \bar{x}_{j,l-\tau})}{2 \sum_{\tau=1}^{Q} \tau^2}, j = 0, 1, \ldots, D, \quad l = 1, 2, \ldots, T. \tag{5.11}$$

The same concept can be applied when computing the second temporal derivatives $\Delta^2 \bar{x}_{j,l}$ (aka delta–delta coefficients). The second temporal derivatives are based on the values $\Delta \bar{x}_{j,l-\tau}$ and $\Delta \bar{x}_{j,l+\tau}$ computed as in eq. (5.11). Therefore, the second temporal derivative, $\Delta^2 \bar{x}_{j,l}$, can be computed as

$$\Delta^2 \bar{x}_{j,l} = \frac{\sum_{\tau=1}^{Q} \tau (\Delta \bar{x}_{j,l+\tau} - \Delta \bar{x}_{j,l-\tau})}{2 \sum_{\tau=1}^{Q} \tau^2}, j = 0, 1, \ldots, D, \quad l = 1, 2, \ldots, T. \tag{5.12}$$

The first and second temporal derivatives are often appended to the static audio descriptors, which results in a feature vector with double or triple the number of audio descriptors. Feature selection could be carried out to eliminate the less relevant descriptors.

Including the third, $\Delta^3 \bar{x}_{j,l}$, and the higher temporal derivatives in the audio feature vector are also feasible; however, these are rarely used and have not been unambiguously proven advantageous in practical applications yet.

5.3.3 Shifted delta coefficients

The shifted delta coefficients (SDCs) aim to capture certain long-term temporal dependences present in a sequence of audio features. This is implemented by staking together blocks of delta coefficients that are spaced at significant distance P, where P is typically much bigger than the selected size of Q in eq. (5.11).

In the case of simple first temporal difference, we can compute the delta coefficients (5.11) multiple times over the sequence of audio feature vectors with some step P (Figure 5.3). Here, P is an integer that specifies the skip step size in terms of l segments. The k blocks of delta coefficients computed this way are staked together to form a new vector of SDC.

The size and the values of a SDC vector depend on three adjustable parameters: Q, P, and k. Here, Q is the number of audio feature vectors considered in the computation of delta coefficients, P is the step specifying the spacing between subsequent blocks, and k is the desired number of blocks which are stacked together to form the desired SDC vector.

Based on the same input feature vector $\bar{x}_j$ and selecting different values of P and k, we can obtain a number of SDC vectors which have a different temporal resolution:

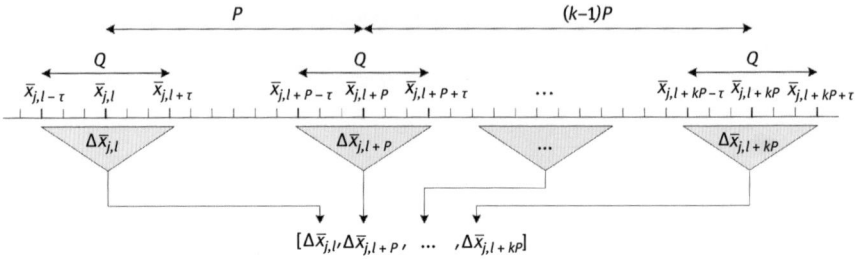

Figure 5.3: Computation of the shifted delta coefficients feature vector.

$$\Delta \bar{x}_{j,i} = \bar{x}_{j,l+iP+\tau} - \bar{x}_{j,l+iP-\tau}, \qquad (5.13)$$

where $j = 0, 1, \ldots, D$, $l = 1, 2, \ldots, T$, $\tau = 1, 2, \ldots, Q$, $i = 1, 2, \ldots, k$. It is possible that the delta coefficients $\Delta \bar{x}_{j,l}$ are computed through a subset of Q feature vectors, depending on the selected values of the respective weights.

Concluding remarks

The post-processing discussed here attempts to compensate partially for the main weakness of 1-D audio parameterization methods – during audio parameterization, they do not take advantage of any temporal information that might be available in a sequence of audio segments.

In Chapter 6, we extend the assortment of audio parameterization methods considered here with a number of recent 2-D methods, which already gained momentum in research on automated acoustic recognition of sound-emitting species and in unsupervised clustering of audio.

Collectively, the 1-D and 2-D audio feature extraction methods constitute the basis of contemporary audio parameterization. Their use depends on the specifics of the applications, but also on their advantages and limitations. In this regard, the most significant advantage of the 1-D audio parameterization methods is that they can be implemented to operate online, in real time, or near real time. This means that all processing of audio and computation of descriptors is completed before the next block of audio is acquired or with a very small delay with respect to the present time moment. The 2-D methods typically operate offline with pre-recorded files. This is because the 2-D methods process multiple audio segments simultaneously, and therefore require a larger buffering of the audio signal, which would mean a significant delay with respect to the present moment.

6 Two-dimensional audio parameterization

Introduction

A new paradigm for audio pre-processing and parameterization, which makes use of image processing techniques, was recently introduced. A common feature of all methods grounded on this paradigm is that these compute the audio spectrogram, which is then considered a greyscale image, and image processing techniques are applied in order to implement noise suppression, audio parameterization, or both. These methods always operate simultaneously on multiple columns and rows of the spectrogram, that is, take advantage of the concurrence of temporal and frequency information and by that reason are referred to as 2-D audio parameterization methods. The simultaneous operation along the temporal and frequency dimensions in the time–frequency (aka spectra-temporal) domain provides certain advantage. This advantage is due to the opportunity (i) for selective suppression or enhancement of specific signal content, and, more importantly, (ii) of handling entire sound events as well-defined 2-D objects. When compared to 1-D methods, these new functionalities contribute towards an improved suppression of background noise and specific types of interference, and also help in the reliable identification of regions of interest (ROI) in the spectrogram that correspond to potentially important sound events.

The price paid for such simultaneous operation along the temporal and frequency dimensions is that the outcome of audio processing might (and typically will) depend on future values of the signal, with respect to the time moment for which the output is computed. The last does not permit for the real-time implementation of these algorithms. Since audio is typically processed in large blocks of samples, the near-real-time or the offline processing of sound recordings is commonly used anyway. However, the 2-D audio processing significantly increases this delay. Therefore, the 2-D methods are particularly appropriate for applications where real-time processing of sound recordings is not mandatory. Among these are many computational bioacoustic tasks oriented towards the development of tools for species or sound events detection, identification, and recognition, contents extraction, and so on.

The main goal of 2-D audio parameterization methods is to obtain a compact set of descriptors which capture the statistical properties of the underlying sound source based on the observed time–frequency distribution of energy in the automatically selected ROI. The signal descriptors are subsequently fed to state-of-the-art statistical classification algorithms or similarity search techniques in order to recognize a set of animal sounds or to cluster data with respect to a certain predefined criteria (Bardeli 2009; Briggs et al. 2012; Aide et al. 2013; Kaewtip et al. 2013; Lee et al. 2013; Potamitis 2014; Ventura et al. 2015). In the following sections, we overview the basic signal processing steps in several indicative 2-D audio parameterization methods.

6.1 The audio spectrogram

The audio spectrogram provides a human-friendly representation of the time–frequency distribution of energy in a sound. In brief, the audio spectrogram is usually presented as a 2-D plot, where the abscissa axis stands for time and the ordinate for frequency. The intensity of sound pressure is represented either by the degree of ink darkness or by the degree of brightness of pixels, depending on whether the spectrogram is drawn on a paper or it is observed as a digital image.

The systematic sub-band analysis of audio signals is already in practice since a century ago (Miller 1916; Crandall 1917). Subsequently, the theoretical advances in the area of time–frequency representation of signals[1] (Gabor 1946; Cooley and Tukey 1965) opened new opportunities for the systematic analysis of sounds. This is mainly due to the wider availability of electrical audio spectrographs in the 1950s and 1960s, that is, equipment capable of drawing on paper greyscale audio spectrograms. Since then, the analysis of audio spectrogram is considered as an essential instrument in bioacoustic studies. The recent advances in information technology and the greater computing power of contemporary microprocessors made the fast calculation of high-resolution spectrograms, their automated processing, and the automated estimation of a large number of parameters describing the sound source characteristics solely based on the observed sound events feasible.

In order to reduce duplications in the exposition and to facilitate the presentation of the 2-D audio processing methods in the next sections, here we formalize the audio pre-processing steps and the spectrogram computation. These are assumed common to all methods discussed in Chapter 6, unless explicitly stated otherwise.

In brief, as discussed in Chapter 5, the audio pre-processing steps aim to (i) adjust the sampling frequency for the requirements of the particular application set-up, (ii) implement some noise reduction strategy, and (iii) segment the audio recordings to chunks with duration of few seconds and short-term stationary segments in the range of milliseconds. Under the assumption that

---

[1] A brief historical note about the roots of the fast Fourier transform is available in Cooley et al. (1967).

the audio recordings are captured with a digital device, and therefore the signal is already available in pulse code modulation format, the pre-processing that follows includes the following:

1. Resampling the audio signal, which most often means downsampling. Downsampling helps to reduce the computational and memory demands in the next signal processing steps.
2. High-pass filtering of the downsampled signal, aiming to reduce the influence of low-frequency noise from the environment, due to vibrations of equipment, direct wind blows in the microphone, vegetation movements due to wind, and other low-frequency interferences, which are outside of the frequency range of interest.

Afterwards, the dynamic spectrum of the pre-processed discrete time domain signal $s(n)$ is computed by the short-time discrete Fourier transform (STDFT):

$$S(k,l) = \sum_{n=0}^{N-1} s(n) W(n+lL) \exp\left(-\frac{j2\pi nk}{N}\right), 0 \leq n, k \leq N-1, \quad (6.1)$$

where n is the index of the time domain samples, k is the index of the Fourier coefficients, and l denotes the relative displacement of the current audio segment in terms of steps of L samples. Here, $W(m)$ is a certain window function. Although any window function is applicable (Harris 1978), the most common choice is the Hamming window because it provides a good trade-off between time and frequency domain resolution. The Hamming window, $W(m)$, defined as

$$W(m) = 0.54 - 0.46 \cos\left(\frac{2\pi m}{M}\right), m = 0, 1, \ldots, M-1, \quad (6.2)$$

is applied to reduce the spectral distortions caused by an abrupt change of signal amplitude at the boundary points of each audio segment.

Here, the STDFT is obtained after applying the discrete Fourier transform (DFT) for N samples on the zero-padded signal $s(n)$ weighted with a sliding window function of M samples. The Hamming window is sliding with a step of L samples between subsequent segments.

The spectrogram $|S(k,l)|^p$ obtained as the power p of the absolute dynamic spectrum (6.1) is considered a greyscale image and is the main source of information in the subsequent image processing steps. The superscript p is a positive real number. Most often, the spectrogram is defined for $p = 2$, so that we can interpret it as the distribution of energy over the time–frequency plane. However, values $1 \leq p \leq 2$ are also feasible options as the choice of p does not affect the time–frequency localization of sound events but only changes the dynamic range of the spectrogram magnitudes, and thus the contrast between small and large

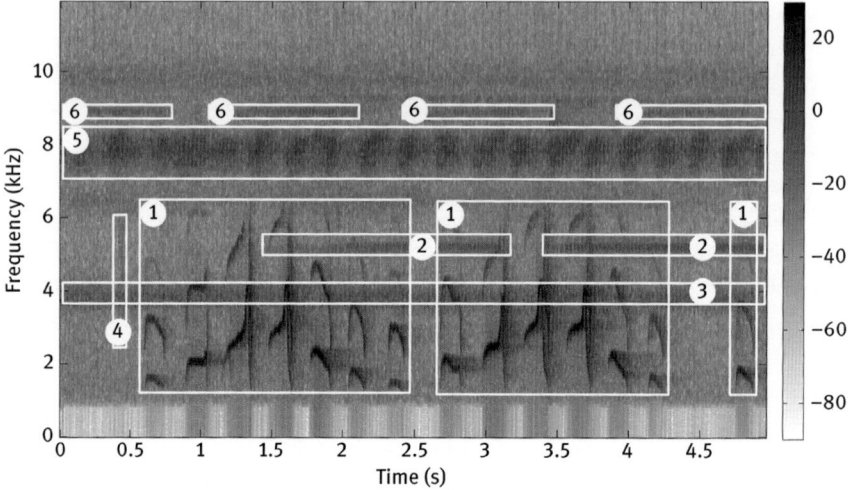

Figure 6.1: The audio spectrogram for a 5-s chunk of real-world recording, after the signal is downsampled to 24 kHz. A high-pass filtering with cut-off frequency 1,000 Hz was applied.

magnitudes. The magnitudes in the spectrogram are often represented in a logarithmic scale. Despite the preferred magnitude compression law, it is important that the amplitudes in the spectrogram remain linked in a reasonable manner to the intensity of the original input audio signal.

For convenience of explanation in the following discussions, let us consider the audio spectrogram $|S(k,l)|$ shown in Figure 6.1, which was computed for a 5-s excerpt[2] of a real-world recording. Please note that the specific audio chunk is characterized with a relatively high SNR and contains multiple sound events of six different species. This audio chunk was downsampled to 24 kHz and then was high-pass-filtered with a Butterwort filter of order 10, according to the preprocessing steps stated above. Here, the cut-off frequency $f_{cut} = 1,000$ Hz of the high-pass filter was arbitrary chosen. The choice of appropriate cut-off frequency is an application-dependent issue, and thus, values could vary in a wide range.

In order to facilitate the discussion on methods and the interpretation of results presented in Sections 6.2–6.6, the audio spectrogram has been manually annotated on the level of sound events. Specifically, the sound events tagged with

---

[2] A 5-s excerpt from recording MYRMUL09.mp3 (offset 13 s from the start). The dominant species tagged with /1/ in Figure 6.1 is the Amazonian Streaked Antwren (*Myrmotherula multostriata*), recorded by Jeremy Minns, 2003-04-10, 15:58h, Brazil, Pousada Thaimaçu. Rio São Benedito. Source: Xeno-Canto. Available online at http://www.xeno-canto.org/sounds/uploaded/DGVLLRYDXS/MYRMUL09.mp3

/1/ belong to the bird species Amazonian Streaked Antwren (*Myrmotherula multostriata*), which here we regard as the target species. The sound events with tags /2/-/6/ originate from other species, which share the same habitat.

Given that we are interested in extracting audio descriptors for the sound events associated with tag /1/, we need to address successfully two key issues:
1. There is an overlap between the sound events of the target species /1/ with these of other species.
2. Some sound events of other species (e. g. /5/) have momentous energy higher than one of the target species /1/.

Indeed, the sound events tagged with /2/ and /3/ overlap in time and in frequency range with the sound events of the target species /1/. Therefore, any noise suppression method, which is based on spectral subtraction in the frequency domain or on filtering in the time domain, will not be effective – it might reduce the interference but will alter the spectrum of the sound events of the target species as well. This is quite a challenging problem. Furthermore, another difficulty is that the sound event tagged with /5/ has a relatively high short-time energy, which is higher than the weak parts of the sound events tagged with /1/, and therefore, the commonly used energy-based thresholding methods will not lead to an effective interference reduction.

Therefore, some automated selective filtering approach is required in order to remove any interference (e. g. /2/-/6/), without altering the target sound events /1/. In the following subsection, we evaluate the applicability of the spectrogram thresholding approach, and then we will discuss some methods that implement selective filtering of the spectrogram.

6.2 The thresholded spectrogram

As real-world recordings usually contain plenty of environmental noise and competing sounds of multiple species, the quality of the raw spectrogram is not sufficient to perform an accurate selection of the desired high-energy areas of interest or to compute reliable audio descriptors. A simple method for reducing the influence of environmental noise in the spectrogram is to reset to zero all frequency components with amplitude under a certain threshold θ:

$$S_\theta(k,l) = \begin{cases} |S(k,l)| & \text{if } \theta_k \geq |S(k,l)| \\ 0 & \text{otherwise} \end{cases}, 0 \leq k \leq N-1, 0 \leq l \leq T-1, \quad (6.3)$$

where k is the frequency bin index, l is the index of the audio segment, N is the DFT size, and T is the total number of audio segments in the spectrogram. The threshold θ is usually set proportional to the mean value, computed over the frequency and time dimensions of the entire spectrogram:

$$\theta = \frac{1}{T}\sum_{l=0}^{T-1}\left(\frac{1}{N}\sum_{k=0}^{N-1}|S(k,l)|\right), 0 \le k \le N-1, 0 \le l \le T-1. \quad (6.4)$$

Alternatively, an individual threshold θ_k could be computed for each frequency band k as the mean value over the entire duration of the spectrogram:

$$\theta_k = \frac{1}{T}\sum_{l=0}^{T-1}|S(k,l)|, 0 \le k \le N-1, 0 \le l \le T-1. \quad (6.5)$$

In Figure 6.2, we show the outcome of applying eq. (6.3) to the spectrogram in Figure 6.1. All frequency components $|S(k,l)|$, which are smaller than the mean of each frequency band θ_k, were reset to value zero. (In Figure 6.2, we use logarithmic representation of the spectrogram magnitudes, and zero is represented as -100 dB.)

As shown in Figure 6.2, certain amount of background noise has been removed together with the sound events tagged with /3/ and /4/. However, the processed spectrogram still contains some residual noise and interferences overlapping with the sound events of category /1/. Some fragments of the sound events

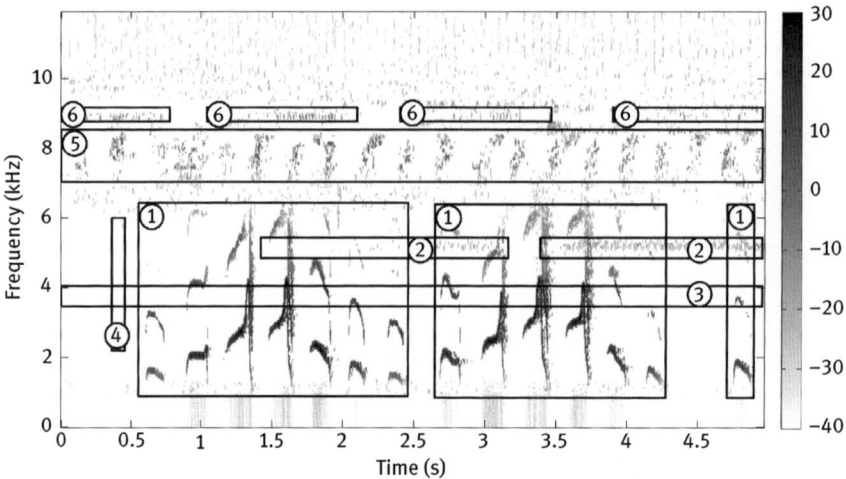

Figure 6.2: Spectrogram of the audio signal after thresholding by magnitude.

with tags /2/, /5/, and /6/ are with higher amplitude than the amplitude of details in /1/, and thus cannot be eliminated via simple thresholding procedure without degrading the target sound events /1/. By that reason, the spectrogram thresholding method is rarely used alone; however, it is often used as an important processing step in multistage methods.

6.3 Morphological filtering of the thresholded spectrogram

Morphological filtering (Heijmans and Ronse 1990) is a well-known image processing technique which allows selective filtering of an image depending on the structuring element size and shape. Since the outcome of the operations erosion and dilation largely depends both on the image contents and on the choice of size and shape for the structuring element, usually the last is defined based on some prior knowledge about the 2-D objects we would like to eliminate, preserve, model, or recognize.

Recently, morphological image processing techniques (Soille 1999; Najman and Talbot 2010) made their way into audio processing tasks, primarily in applications where offline processing of audio recordings is an acceptable option. This happened mainly because the morphological image processing techniques provide the means for dealing with environmental noise and interferences in a new flexible manner, which complements traditional 1-D signal processing methods.

The incorporation of prior knowledge about the sound events of interest allows proper adjustments of the morphological filtering process, and therefore it could be used to optimize the separation of sound event of different sources (cf. Figure 6.1). In such a way, the use of prior knowledge may directly increase the probability of sound events originating from different sources to be separated, and therefore considered as different acoustic events during the automated recognition algorithms. The merging of sound events originating from different species is not desirable and gives rise to increased inaccuracy during the process of modelling and recognition of sound events, especially when statistical machine learning methods are used in the single-label recognition tasks.

In Figure 6.3, we show the overall block diagram of a more advanced spectrogram processing method, which is based on morphological filtering of the thresholded spectrogram. The audio pre-processing steps, the computations of the dynamic spectrum, and the spectrogram follow the exposition in Section 6.1. However, the method described here implements additional image processing steps applied on the thresholded spectrogram (Figure 6.2) in order to suppress the residual noise after the thresholding through eq. (6.3).

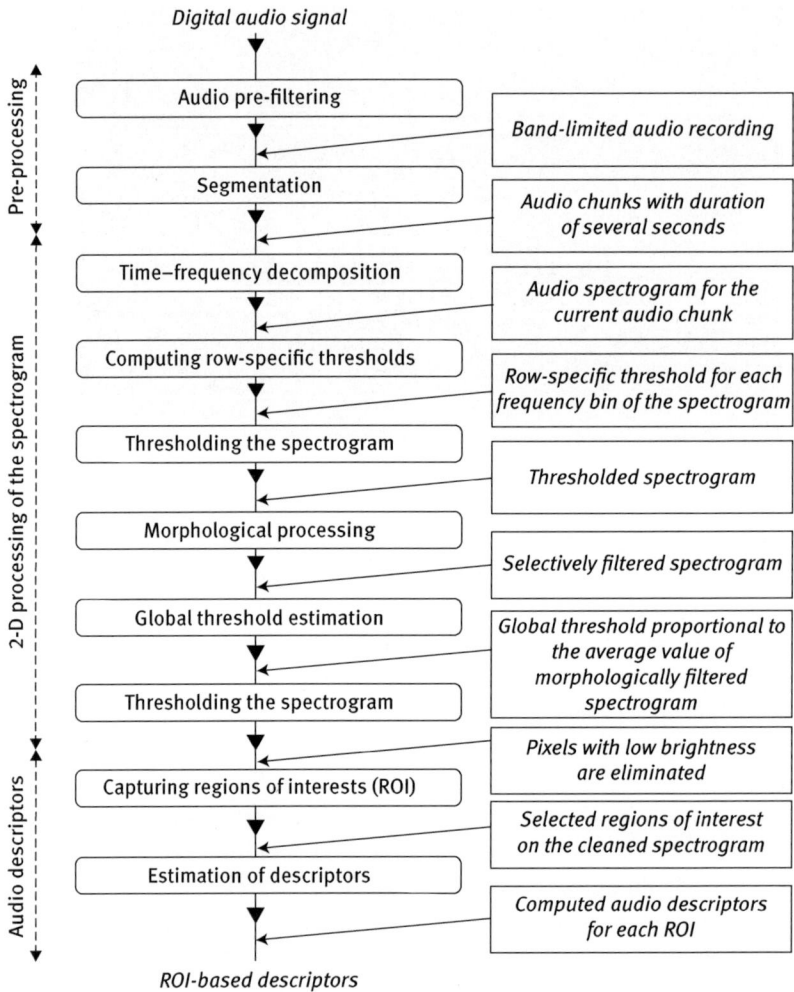

Figure 6.3: Audio parameterization with 2-D processing of the spectrogram and ROI-based features.

In order to avoid unnecessary reiteration of explanations, we outline the method starting with the application of eq. (6.3) on the spectrogram $|S(k,l)|$, computed as the absolute value of the real part of the dynamic spectrum (6.1). The outcome is the thresholded spectrogram presented in Figure 6.2, computed as

$$S_\theta(k,l) = \begin{cases} |S(k,l)| & \text{if } \theta_k \geq |S(k,l)| \\ 0 & \text{otherwise} \end{cases}, 0 \leq k \leq N-1, 0 \leq l \leq T-1. \quad (6.6)$$

Here, k and l are the indices for the frequency bin and the number of the audio segment, N is the DFT size, and T is the total number of audio segments in the spectrogram. The threshold θ_k is computed for each frequency band k as the mean value over the entire duration of the spectrogram:

$$\theta_k = a\frac{1}{T}\sum_{l=0}^{T-1} |S(k,l)|, 0 \le k \le N-1, 0 \le l \le T-1 \tag{6.7}$$

The parameter a allows fine-tuning of the decision threshold θ_k so that the spectrogram thresholding is optimized with respect to a certain cost function. The parameter a could be set based on prior knowledge, heuristically, or alternatively can be adjusted automatically – for instance, via particle swarm optimization (Parsopoulos and Vrahatis 2010), differential evolution (Storn and Price 1997), or some gradient-based method for optimization (Polak 1997). The automated optimization procedure depends on the availability of representative development data set of few dozen spectrograms, which are used in the adjustment of a. The cost function used in the optimization algorithm should seek a certain predefined trade-off between maximization of the energy of the target sound event (in our example tagged with /1/) and minimization of the energy of the non-target sound events (in our example tagged with /2/, /3/, /4/, /5/, and /6/).

After applying the operator *erosion* on the thresholded spectrogram (6.6), we obtain a spectrogram cleaned from details smaller than the selected structuring element C:

$$S_{\theta e}(k,l) = S_\theta(k,l) \odot C = \{z : C_z \subseteq S_\theta(k,l)\}. \tag{6.8}$$

In the present example, we consider the structuring element C as a square with dimensions 7×7 pixels. The shape and size of the structuring element is an application-specific issue and does not bind the method to the illustrative example considered here. The outcome of applying eq. (6.8) on the thresholded spectrogram (Figure 6.2) is shown in Figure 6.4. As shown in the figure, the operator *erosion* shrinks the bright objects, that is, the regions with high energy, and eliminates from the spectrogram the unimportant blobs whose dimensions are smaller than the size of the structuring element C.

Comparing the images shown in Figures 6.2 and 6.4, we observe that the cleaned spectrogram in Figure 6.4 is more or less free of any competing sound events, including these with tag /5/, which are well seen and prominent in the image in Figure 6.2. The sound events tagged with /1/ were degraded to some extent;

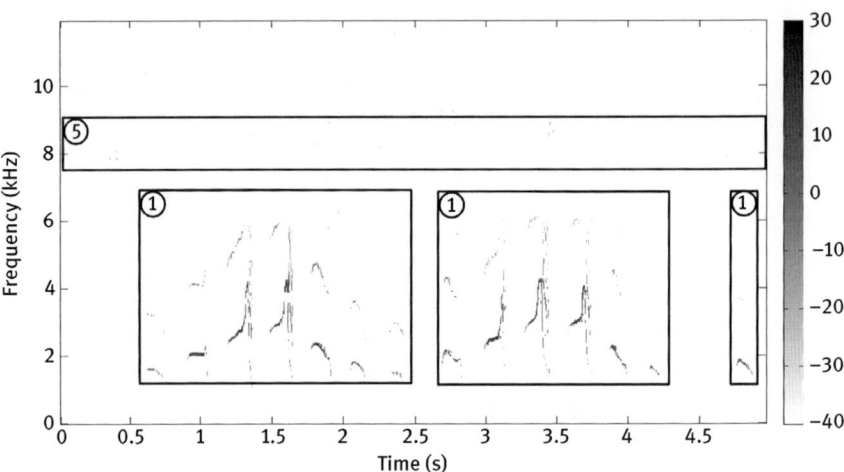

Figure 6.4: The thresholded spectrogram after applying the erosion operator.

however, these are still clearly discernible in the image, which indicates that most of the important information about these vocalizations was kept intact.

A well-established approach for recovering eroded contours is the dilation operator, which fills in the missing patches in the structure of an object. Applying the dilation operator on the spectrogram eroded through eq. (6.8), using the same structuring element C, would recover the loss of quantity in the objects, which survived the erosion operator. Therefore, the dilated spectrogram

$$S_{\theta ed}(k,l) = S_{\theta e}(k,l) \oplus \hat{C} = \{z : \hat{C}_z \cap S_{\theta e}(k,l) \neq \emptyset\} \quad (6.9)$$

would have a smaller number of objects than the eroded spectrogram (Figure 6.4). These objects however would generally have larger dimensions due to the quantitative enlargement depending on the shape and size of the structuring element $\hat{C}$. Therefore, applying the dilation operator (6.9) on the eroded spectrogram results in a smaller number of objects due to the merging of objects located in a close proximity, that is, when the gap between objects is smaller than the dimensions of the structuring element $\hat{C}$.

In the general case, the shape and size of the structuring element $\hat{C}$ are not bound by the choice of the structuring element C in eq. (6.8). Whether the two structuring elements shall be selected identical or not is an application-specific issue, which depends on whether we only wish to recover the pixels eroded through eq. (6.8), or we also intend to smooth or restructure the objects of interest in the spectrogram.

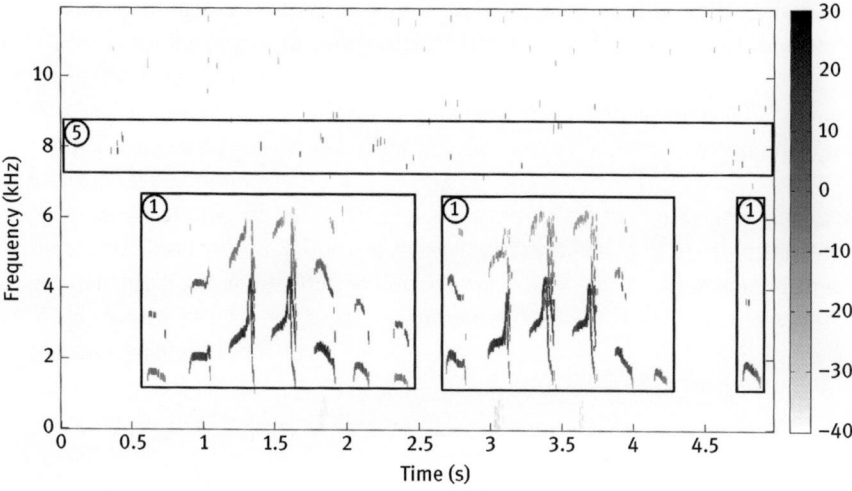

Figure 6.5: The spectrogram after applying the dilation operator.

The outcome of applying eq. (6.9) with a structuring element $\hat{C} \equiv C$ on the eroded spectrogram (Figure 6.4) is presented in Figure 6.5. As shown in the figure, the residual noise that survived the erosion operator (6.8) was amplified after applying the dilation operator (6.9) – this is an undesired side effect. However, as shown in Figure 6.5 the main purpose of the dilation operator is fulfilled successfully – all objects belonging to the sound events tagged with /1/ were recovered completely, and nearly all interference from competing sound events were removed. Exceptions are some miniature residual blobs in the area of tag /5/. Their existence is conditioned on the fine-tuning of the threshold (6.7) and on the size and shape of the structuring elements in eqs. (6.8) and (6.9). Optimal selection of these free parameters of the algorithm would guarantee minimum residual noise in the morphologically filtered spectrogram.

In the following processing steps, we deal with the residual noise and attempt to reconstruct the original objects in the morphologically filtered spectrogram (6.9). For that purpose, we implement the following post-processing steps:

1. Estimate the decision threshold θ_{ed}, proportional to the mean value of all pixels in the morphologically filtered spectrogram.
2. Apply the threshold on the morphologically filtered spectrogram by preserving the elements with energy above the threshold and resetting to zero those below the threshold.
3. Replace the values of the preserved elements with these from the original spectrogram $|S(k,l)|$.

In brief, the decision threshold θ_{ed} is computed as proportional to the mean value of all frequency elements in the morphologically filtered spectrogram:

$$\theta_{ed} = A \frac{1}{T} \sum_{l=0}^{T-1} \left(\frac{1}{N} \sum_{k=0}^{N-1} S_{\theta ed}(k,l) \right), 0 \le k \le N-1, 0 \le l \le T-1. \qquad (6.10)$$

Similarly to eq. (6.6), k and l correspond to the frequency bin and audio segment indices, N is the DFT size, and T is the total number of audio segments in the morphologically filtered spectrogram. Here, A is a parameter which could be selected empirically or adjusted automatically based on development dataset. Any value of the parameter A in the range [1, 20] usually provides good results. Although the value of A is not crucial for the proper operation of the method, it also can be adjusted automatically with the use of certain genetic or gradient-based optimization algorithm and a cost function that provides some balance between distortion of the target event and the energy of residual noise.

Once the value of the threshold θ_{ed} is selected, we can implement the thresholding and the replacement of values according to steps (2) and (3), as follows:

$$S_{th}(k,l) = \begin{cases} |S(k,l)| & \text{if } \theta_{ed} \ge S_{\theta ed}(k,l) \\ 0 & \text{otherwise} \end{cases}, 0 \le k \le N-1, 0 \le l \le T-1. \qquad (6.11)$$

The outcome of eq. (6.11) is a spectrogram (cf. Figure 6.6) that preserves the target sound events nearly identical to those in the original spectrogram but still takes advantages of the morphological filtering and the post-processing steps that remove competing sound events.

Although the selective morphological filtering method discussed here makes use of some prior knowledge about the target sound events, which is incorporated mainly through the settings of the morphological filtering, such as the size and shape of the structuring elements, this method is applicable to a wide range of sound-emitting species.

For some audio parameterization methods, the preservation of the original properties of sound events (both their location, shape, and energy) is of crucial importance because on this depends the representativeness of the computed audio descriptors. This holds true particularly strongly for the computation of Mel-frequency cepstral coefficients (MFCC) parameters and for other audio parameterization methods that model the temporal evolution of energy in the frequency bands of interest. However, the use of the selective morphological filtering method discussed here results in a spectrogram with many zeros in the columns

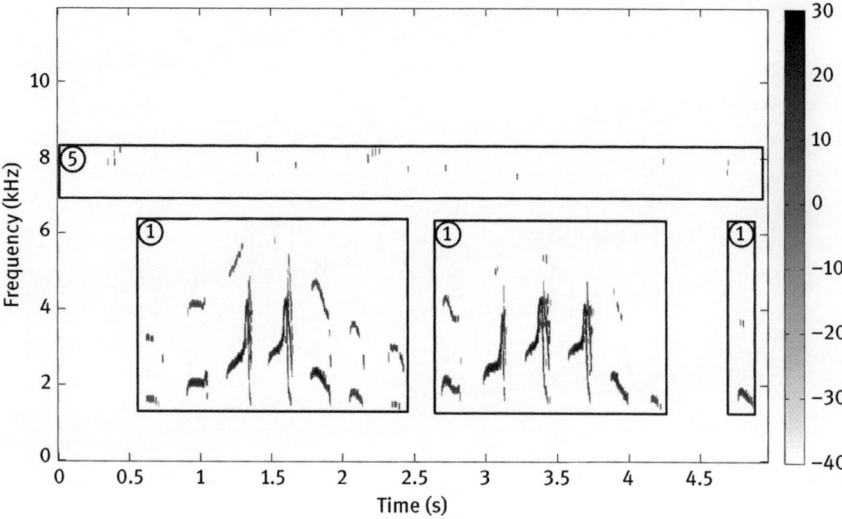

Figure 6.6: Spectrogram after applying the threshold on the morphologically filtered spectrogram.

and rows, and therefore MFCC would not lead to an efficient representation of sound events. However, some 2-D audio parameterization methods do not depend on the preservation of all information about the original sound events, and therefore, a significant portion of the information could be eliminated by decreasing the overall complexity and memory usage. For instance, this applies to the case where the audio feature extraction stage takes advantage only of the 2-D contours of the sound events.

The morphological filtering method presented in this section is not tuned to the computation of a specific set of audio descriptors. Instead, it could serve as a pre-processor for different audio parameterization algorithms, among which are cepstral coefficients, wavelet packet-based descriptors, or other 1-D audio descriptors. Furthermore, the method is compatible with the estimation of image-based descriptors based on ROI, which aim to discover 2-D shapes and contours in the cleaned spectrogram. For instance, based on the outcome of eq. (6.11), it is straightforward to estimate the MFCC (cf. Section 6.4, as defined starting from eq. (6.17) onwards). In the same time, the outcome of applying eq. (6.11) provides the means for computing image-based audio descriptors based on 2-D ROI (cf. Section 6.6) or for computing multidimensional audio descriptors through structure tensor-based detection of potentially interesting locations in the spectrogram (cf. Section 6.5).

6.4 MFCC estimation with robust frames selection

As already said above, a central challenge in the analysis of real-world audio recordings is the presence of noise and competing sounds of multiple species. Therefore, methods that aim to compute reliable audio descriptors need to reduce noise, and if possible, to preselect the most promising portions of the signal. It is expected that the latter would significantly improve the chances of successful modelling and recognition of the target sound categories. As an example of this group of methods, we overview the method of Ventura et al. (2015), which operates on a single-channel soundscape recordings. The main motivation behind this method is to preselect only audio chinks, which correspond to some prominent sound events and ignore sporadic interferences with energy localized in a small time–frequency area. The method studied by Ventura et al. bears some similarity to the method discussed in Section 6.3 because it also makes use of morphological filtering of the spectrogram. In the following, we discuss the main processing steps and explain the differences.

Specifically, Ventura et al. (2015) integrated morphological filtering of the audio spectrogram (considered as an image) and a procedure for robust frame selection into the MFCC computation process. The morphological processing of the spectrogram allows reduction of noise and to some extent elimination of competing sound events with short duration and compact representation in the time–frequency domain. This was reported to facilitate the selection of reliable audio frames for which the MFCC are computed afterwards. The main differences in Ventura et al. (2015) with respect to the pre-processing method discussed in Section 6.3 are as follows:

1. The method discussed in Section 6.3 operates on the thresholded spectrogram, while Ventura et al. implement morphological filtering directly on the original spectrogram (Figure 6.1).
2. The method of Ventura et al. uses a different set-up of the morphological filtering and allows much larger structuring elements to be used, which provides more flexibility to deal with interferences.
3. The method of Ventura et al. makes use of histogram-based algorithm for robust frame selection, while no frame selection is available in the method described in Section 6.3.
4. The method of Ventura et al. selects entire audio frames directly from the original audio signal in order to compute audio features, while the method in Section 6.3 assumes that sound event descriptors will be computed based on the cleaned audio spectrogram (the last means partial distortion of target sound events due to the morphological processing).
5. The method of Ventura et al. aims to compute MFCC audio descriptors for the selected audio frames, while the method discussed in Section 6.3 is

more general and allows various audio descriptors to be computed based on the cleaned spectrogram. Exceptions are cepstral coefficient-like descriptors (e. g. linear-frequency cepstral coefficients [LFCC], MFCC, Greenwood function cepstral coefficients [GFCC], etc.), for which the method discussed in Section 6.3 is not suitable due to the large areas of the spectrogram forced to value zero.

These differences chart the area of applicability of the two methods depending on the requirements of the specific application task. In brief, Ventura et al. (2015) addressed this challenge by an audio parameterization method, which makes use of morphological filtering applied directly on the audio spectrogram considered as an image, that is, without prior thresholding. There, the morphological processing aims at selective filtering of the spectrogram so that sound events, which are smaller than the dimensions of the target sound events, are eliminated (Figure 6.7). The selectively filtered spectrogram is used as the basis for robust frame selection that eliminates the less promising portions of the signal. The morphological filtering and the robust frame selection were incorporated in the computation of MFCC and advantageous performance and speed up of computations with respect to other methods were reported.

The morphological operators offer a convenient way of image enhancement by noise suppression and simplification of the spectrogram by retaining only those components that satisfy certain size or contrast criteria. Therefore, in order to reduce the environmental noise, Ventura et al. (2015) made use of the morphological processing on the audio spectrogram by means of two morphological operators (Bovik 2005). Similarly to Cadore et al. (2011) and Potamitis (2014), Ventura et al. (2015) apply the opening morphological operator, which consists of the erosion operator (6.12), followed by the dilation operator (6.13).

$$S_e(k,l) = |S(k,l)| \odot B = \{z : B_z \subseteq |S(k,l)|\}, \quad (6.12)$$

$$S_{ed}(k,l) = S_e(k,l) \oplus \hat{B} = \{z : \hat{B}_z \cap S_e(k,l) \neq \emptyset\}. \quad (6.13)$$

Here, $|S(k,l)|$, $S_e(k,l)$, and $S_{ed}(k,l)$ are the images corresponding to the spectrogram after the pre-processing step, then after applying the operator erosion, and after applying both the operators erosion and dilation, respectively. In the work of Ventura et al., the structuring element B is a rectangle with dimensions 40×30 pixels. However, the dimensions of the structuring element are an application-specific choice as these depend on the selected sampling frequency, on the overlap among subsequent frames of audio, and the time–frequency footprint of sound events that need to be modelled.

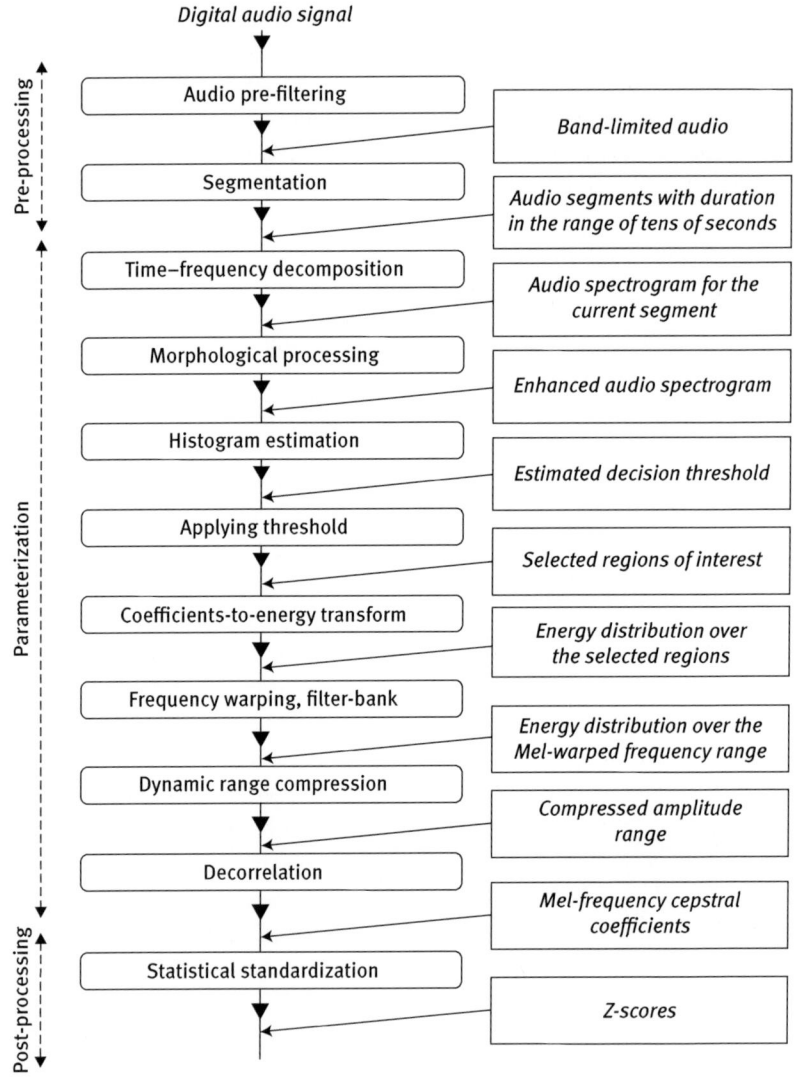

Figure 6.7: Audio parameterization with 2-D processing of the spectrogram and robust frame selection.

The result after applying the operator erosion is the set of all points z such that B translated by z is contained in the image $|S(k,l)|$. In the dilation operator, $\hat{B}$ is the set of pixel locations z, where the reflected structuring element overlaps with foreground pixels in $S_e(k,l)$ when translated to z. The combination of

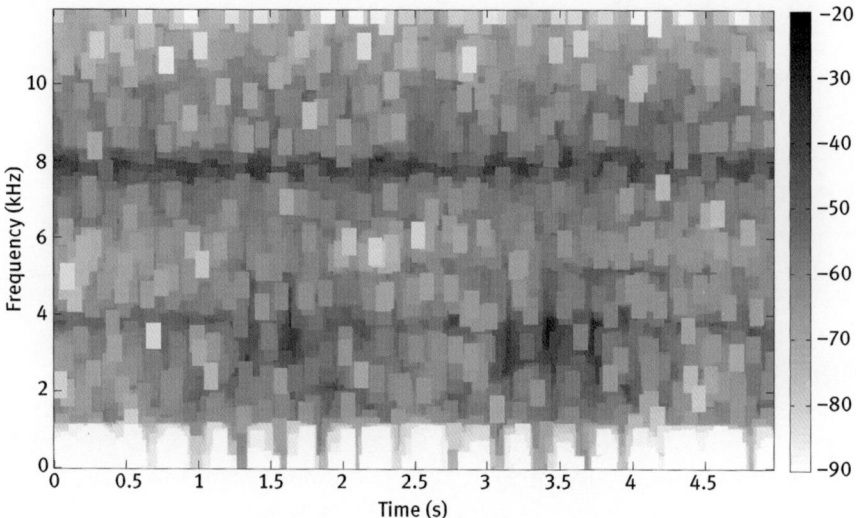

Figure 6.8: The spectrogram after applying the morphological operator erosion.

these two operators defines the operator morphological opening, which can be rewritten as

$$S_{ed}(k,l) = (|S(k,l)| \odot B) \oplus \hat{B}. \tag{6.14}$$

The operator erosion reduces bright regions and enlarges dark regions in the image (cf. Figure 6.8). By contrast, the operator dilation enlarges bright regions and reduces dark regions (Figure 6.9). The subsequent application of both operators eliminates certain small elements in the spectrogram, such as short-lived audio signals and weak distortions in the time–frequency space, such as the sound events /3/ and /4/ in Figure 6.1. The size of the elements that survive this filtering process depends on the size of the structuring elements B and $\hat{B}$.

Comparing the regions of highest energy in Figures 6.8 and 6.9, it is evident that certain elements of the spectrogram that correspond to noise and weak vocalizations of competing sound emissions become less prominent after the use of the operator morphological opening. The resulting spectrogram appears cleaner, and thereby facilitates the more robust and precise selection of the desired audio frames. When the structuring elements B and $\hat{B}$ are identical by size and shape, the order of applying the operators erosion and the dilation could be exchanged. However, in order to obtain similar outcome of the morphological processing, the size and the shape of the structuring elements B and $\hat{B}$ need to be different depending on whether erosion is first or dilation is first.

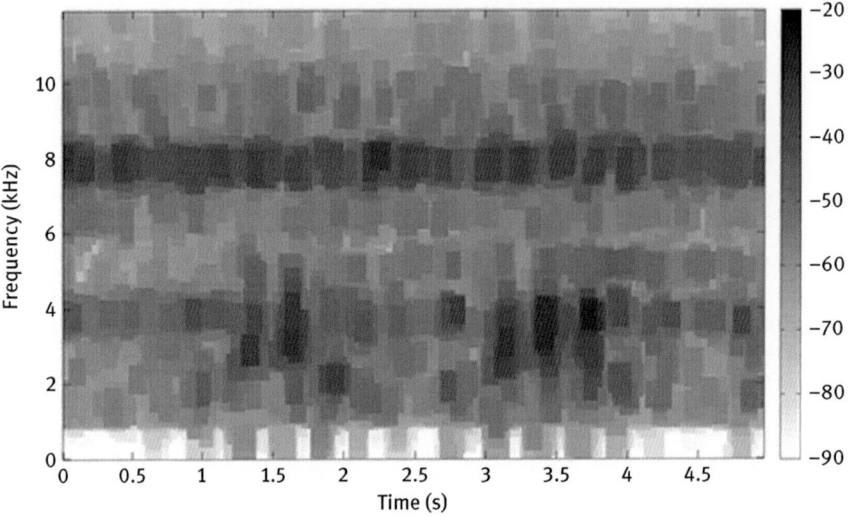

Figure 6.9: The spectrogram after applying the morphological operator dilation.

Robust frame selection is achieved through an algorithm, which adjusts the decision threshold based on the histogram of the sum of the amplitudes for all frequency components in each frame l of the processed spectrogram $S_{ed}(k,l)$:

$$E(l) = \sum_{k=0}^{N-1} S_{ed}(k,l), l = 1, 2, \ldots, T, \quad (6.15)$$

where T is the total number of frames in the specific recording. In the work of Ventura et al. (2015), the threshold θ was determined based on the histogram and equals to the bin in the histogram that contains <30 % of the instances in the highest bin. Only those bins that appear on the right side after the bin with highest value are taken into account. The amplitude value corresponding to the centre of the selected bin specifies the threshold value θ.

The choice of resolution during the computation of the histogram depends on the total number of segments T into the spectrogram. When the spectrogram has several thousands of frames l, an appropriate resolution of the histogram would be in the range [30, 50] bins. The method will also work for a smaller or larger number of bins in the histogram; however, too few bins will introduce significant quantization by level of the energy values represented by the centre of each bin. In such cases, the resolution of the threshold θ will be low (because it will be set with a large discrete), which in turn at the next processing steps might result to elimination of low-energy portions of the target sound events, together with the residual noise. On the other hand, too many bins in the histogram would mean that we might not have enough values in each bin, and therefore will not achieve a smooth distribution of values among subsequent bins. In such case, the threshold θ selection algorithm might be influenced by accidental effects, which may impede the estimation of the threshold value, that is, the value of θ will be biased.

Each frame l, with sum of the amplitudes (6.15) greater than the threshold θ, will be selected for the subsequent processing step. However, please note that Ventura et al. (2015) select frames that are from the original spectrogram $|S(k,l)|$, that is, before the morphological opening is applied:

$$S_{\text{sel}}(k,l_{\text{sel}}) = \begin{cases} |S(k,l)|, & \text{if } E(l) \geq \theta \\ 0, & \text{otherwise} \end{cases}, l = 1, 2, \ldots, T. \qquad (6.16)$$

The audio segments selected from the spectrogram in Figure 6.1 are presented in Figure 6.10. As shown in the spectrogram, most of the frames with significant energy for sound event /1/ were selected for further processing. Eliminating some of the audio frames and robust selection of the audio frames of interest was reported to reduce computational demands during audio parameterization, speed up the pattern recognition process, and reduce the misclassification errors (Ventura et al. 2015).

As shown in Figure 6.10, thanks to the frame selection many audio frames that contain sound events of other species were discarded. The sound events of the target species were preserved due to selective morphological filtering even when

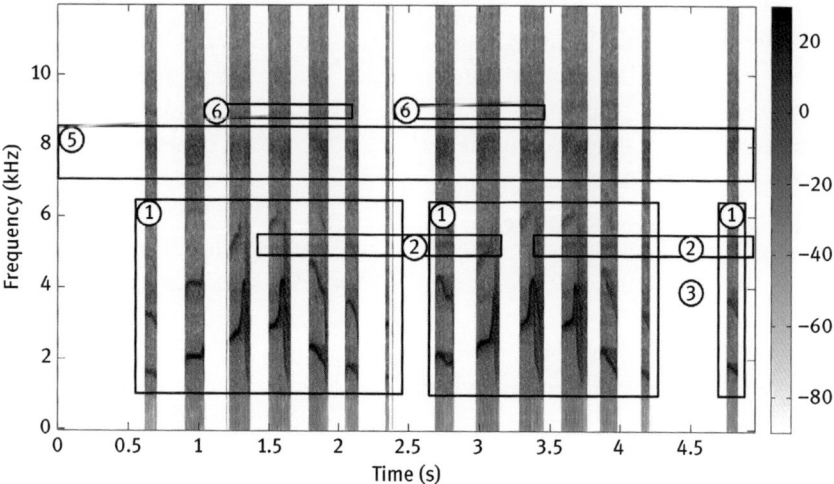

Figure 6.10: Selected frames after applying the algorithm in Ventura et al. (2015) on the spectrogram shown in Figure 6.1.

their amplitude was lower than this of the other co-occurring sounds. Because the selected frames originate from the unprocessed spectrogram in Figure 6.1, the MFCC vectors computed for these frames will represent the mixture of sound events with tags /1/, /2/, /5/, and /6/.

The following processing steps explain the MFCC computation for each frame l_{sel} selected through eq. (6.16). The Mel-scale warping of the frequency scale is implemented through the filter bank $H_i(k)$ consisting of K Mel-spaced equal-height filters acting on the power spectrum $(S_{sel}(k, l_{sel}))^2$ for each selected frame l_{sel}. The log energy for the corresponding filter output is computed as

$$S_m(i, l_{sel}) = \ln\left(\sum_{k=0}^{N-1} (S_{sel}(k, l_{sel}))^2 H_i(k)\right), \quad i = 1, 2, \ldots, K. \tag{6.17}$$

Here, each filter in the filter bank $H_i(k)$ is defined as

$$H_i(k) = \begin{cases} 0 & \text{for } k < f_{b_{i-1}} \\ \frac{(k - f_{b_{i-1}})}{(f_{b_i} - f_{b_{i-1}})} & \text{for } f_{b_{i-1}} \le k \le f_{b_i} \\ \frac{(f_{b_{i+1}} - k)}{(f_{b_{i+1}} - f_{b_i})} & \text{for } f_{b_i} \le k \le f_{b_{i+1}} \\ 0 & \text{for } k > f_{b_{i+1}} \end{cases}, \quad i = 1, 2, \ldots, K, \tag{6.18}$$

where i stands for the ith filter, f_{b_i} are the boundary points of the individual filters, and $k = 1, 2, \ldots, N$ corresponds to the kth coefficient of the N-point DFT. The boundary points f_{b_i} are expressed in terms of position, which depends on the sampling frequency F_s and the number of points N in the DFT:

$$f_{b_i} = \left(\frac{N}{F_s}\right) \cdot \hat{f}_{mel}^{-1}\left(\hat{f}_{mel}(f_{low}) + i \cdot \frac{\hat{f}_{mel}(f_{high}) - \hat{f}_{mel}(f_{low})}{K + 1}\right), \tag{6.19}$$

where the function $\hat{f}_{mel}$ stands for the transformation

$$\hat{f}_{mel} = 1127 \cdot \ln\left(1 + \frac{f_{lin}}{700}\right). \tag{6.20}$$

Here, f_{low} and f_{high} are, respectively, the low and high boundary frequencies for the entire filter bank, K is the total number of filters, and $\hat{f}_{mel}^{-1}$ is the inverse transformation to linear frequency scale, defined as

$$\hat{f}_{mel}^{-1} = f_{lin} = 700\left(\exp\left(\frac{\hat{f}_{mel}}{1127}\right) - 1\right). \tag{6.21}$$

Next, we apply the discrete cosine transform on the result of eq. (6.17) in order to obtain $J + 1$ MFCC parameters:

$$c_j = \sum_{i=1}^{K} |S_m(i, l_{sel})| \cos\left(j(i + 0.5)\frac{\pi}{K}\right), j = 0, 1, \ldots, J. \quad (6.22)$$

Afterwards, the MFCCs are usually standardized for zero mean and unit standard deviation. This results in obtaining the z-score $\bar{c}_j$ of each individual parameter c_j:

$$\bar{c}_j = \frac{c_j - v_j}{\xi_j}, j = 0, 1, \ldots, J. \quad (6.23)$$

The z-scores quantify the distance from the mean value in terms of the standard deviation. The mean value v_j and the standard deviation ξ_j are estimated for each of the $J + 1$ dimensions, for each audio recording separately.

Due to the robust frame selection process outlined above, the MFCC parameters are computed only for the high-energy frames, corresponding to dominant sound events. The last increases the probability for successful recognition of sounds emitted by the target species (sound events tagged with /1/ in Figure 6.1). However, as shown in Figure 6.10, the MFCC vectors computed for the selected frames will represent the mixture of all sound events covered by the filter bank, in the specific case sound events /1/, /2/, /5/, and /6/.

One way to go around this issue is to design a species-specific filter bank, which covers only the frequency range of interest for the target species. However, this will not be practical in the multispecies recognition and species identification set-ups. Another way to deal with this problem is to abandon the traditional MFCC and compute a different set of audio descriptors for each ROI – only for the selected portions of the audio spectrogram. When each ROI coincides with one sound event, one can compute various geometrical descriptors.

6.5 Points of interest-based features

A similarity search framework developed for the needs of information retrieval and classification of animal sounds was outlined in Bardeli (2009). It incorporates a method for points of interest identification and audio parameterization, which makes use of image processing concepts, such as the structure tensor, and image processing techniques applied on the audio spectrogram. Here, we discuss this audio parameterization method not because of its excellence but because we consider it a very interesting research direction, which offers opportunities for advancing further the methods for similarity search in soundscape recordings and some information retrieval concepts relevant to the scope of computational bioacoustics.

The point of interest detection method of Bardeli (2009) relies on the property of structure tensors to integrate information about the orientation of objects at a given location. A structure tensor is used to identify portions of the spectrogram that contain certain 2-D objects and their orientation. The detection of a well-pronounced orientation is assumed an indicator of the presence of a certain sound event at the particular location, while the absence of distinct orientation is associated with silence or random background noise.

The overall block diagram of the audio parameterization method of Bardeli (2009) is illustrated in Figure 6.11. In brief, the audio pre-processing and the audio spectrogram computation are implemented as described in Section 6.1. The audio spectrogram is considered a greyscale image. Next, the 2-D processing of the spectrogram aims to identify the points of interest for which audio descriptors are computed. The procedure for selection of points of interest is based on the use of structure tensor and sub-band processing of the audio spectrogram. The orientation information derived through the structure tensor is used for estimation of an indicator function, referred to as attention value. The attention value is the basis for the subsequent selection of points of interest. In the rest of this section, we focus on the procedure for identification of points of interest in the spectrogram.

Similarly to Sections 6.2–6.4, here we consider the sound spectrogram $|S(k,l)|$ as a greyscale image, which consists of pixels with certain location (k,l) defined in terms of the frequency bin index k and the audio segment index l. Long audio recordings are segmented to chunks of few seconds, and each audio chunk is processed independently of the others. Unlike the discussion in Sections 6.1–6.3, here the brightness of a pixel is defined by the logarithm of the amplitude of sound, that is, the method discussed here operates in the log-magnitude domain.

Let us consider the general case when the dynamic spectrum is computed through eq. (6.1), and the spectrogram is obtained in the log-magnitude domain by logarithmically compressing the magnitude of the absolute dynamic spectrum. Therefore, in the following we actually work with the log-magnitude spectrogram $I(k,l) = \log|S(k,l)|$.

Specifically, interpreting the audio spectrogram as a greyscale image **I**, which is considered a differentiable function of frequency and time $I(k,l)$, we can define the 2-D structure tensor:

$$\mathbf{S}_w = \begin{pmatrix} \mathbf{W}^* \left(I_k(k,l) I_k(k,l) \right) & \mathbf{W}^* \left(I_k(k,l) I_l(k,l) \right) \\ \mathbf{W}^* \left(I_k(k,l) I_l(k,l) \right) & \mathbf{W}^* \left(I_l(k,l) I_l(k,l) \right) \end{pmatrix}, \quad (6.24)$$

6.5 Points of interest-based features — 123

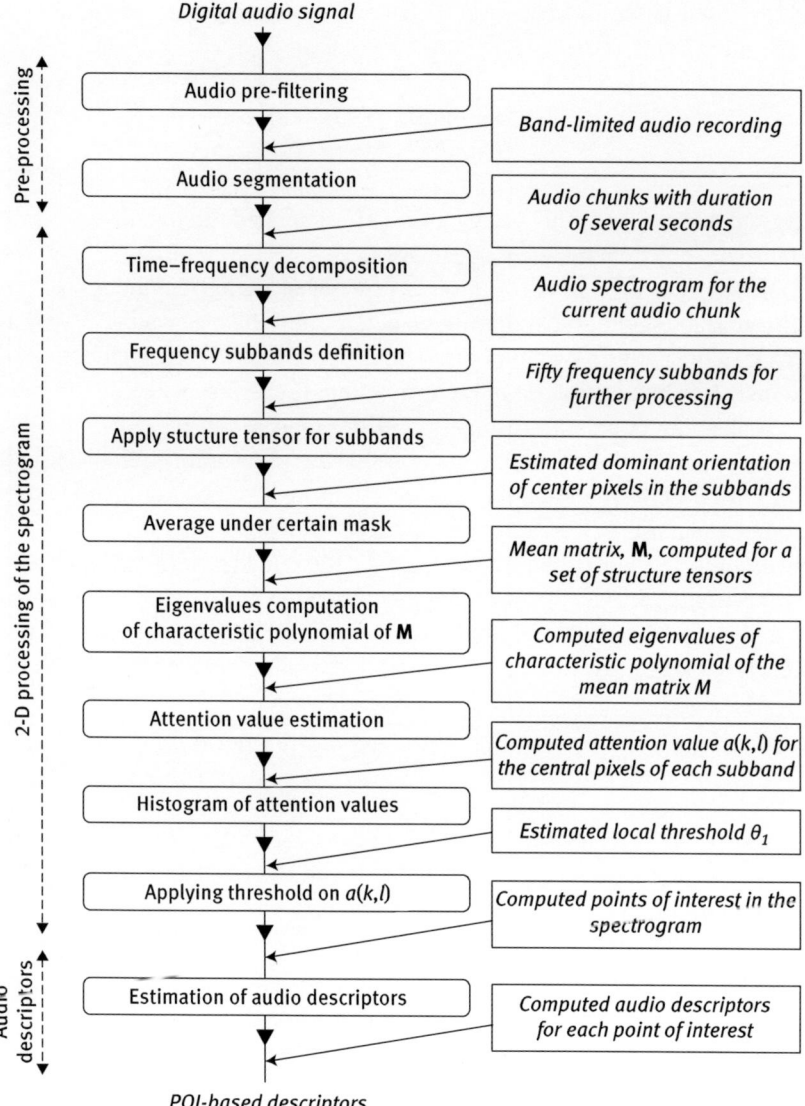

Figure 6.11: The overall block diagram of the audio parameterization method with structure tensor-based estimation of points of interest in the audio spectrogram.

where the smoothing kernel **W** is convolved with the partial derivatives I_k and I_l along dimensions k and l, with

$$I_k(k,l) = \frac{\partial I(k,l)}{\partial k} \text{ and } I_l(k,l) = \frac{\partial I(k,l)}{\partial l}. \tag{6.25}$$

In fact, Bardeli (2009) made use of a structure tensor represented through its matrix-valued array

$$\mathbf{S}_0 = \begin{pmatrix} I_k(k,l)I_k(k,l) & I_k(k,l)I_l(k,l) \\ I_k(k,l)I_l(k,l) & I_l(k,l)I_l(k,l) \end{pmatrix}. \quad (6.26)$$

The partial derivatives I_k and I_l in eq. (6.25) are estimated through the first finite difference directly from the greyscale image $\mathbf{I}$.

The eigenvectors e_1 and e_2 of the structure tensor enclose information about the direction of orientation at a given pixel (k,l), and the eigenvalues λ_1 and λ_2, with $\lambda_1 \geq \lambda_2 \geq 0$, define how pronounced this orientation is. As in $\mathbf{S}_0$, the smoothing kernel $\mathbf{W}$ is reduced to the 2-D unit function. It operates with single pixels, which makes the orientation estimation error-prone in the presence of noise. Therefore, when the method is applied on real-world recordings, the computation of partial derivatives and orientation might be significantly hampered by environmental noise.

In order to deal with this problem, Bardeli (2009) made use of the mean computed for a set of tensors $\mathbf{S}_0$ over a neighbourhood of 11 × 11 pixels. The mean, estimated in each frequency sub-band of the spectrogram $\mathbf{I}$ for a set of structure tensors, is the symmetric matrix $\mathbf{M}$:

$$\mathbf{M} = \begin{pmatrix} A & B \\ B & C \end{pmatrix}, \quad (6.27)$$

which allows a more reliable estimation of the direction and intensity of dominant orientation. Based on the entities A, B, and C, Bardeli (2009) defined the *attention value* $a(k,l)$ to a given pixel (k,l) as a positive number when the discriminant D of the characteristic polynomial of $\mathbf{M}$ is positive. Otherwise, the *attention value* $a(k,l)$ is set to zero, that is,

$$a(k,l) = \begin{cases} \dfrac{A+C}{2} + \sqrt{D}, & \text{when } \lambda_1 > \lambda_2 \\ 0, & \text{when } \lambda_1 = \lambda_2 \end{cases}. \quad (6.28)$$

The discriminant D is defined as

$$D = \left(\dfrac{(A+C)^2}{4}\right) - AC + B^2, \quad (6.29)$$

and the condition the eigenvalues to have different value, $\lambda_1 > \lambda_2$, ensures that the discriminant is larger than zero.

Audio features are extracted only when the *attention value* $a(k,l)$ for pixel (k,l) exceeds a predefined threshold θ_a. The threshold is computed based on a fraction of the histogram of a. For that purpose, the interval $[0, \max_{k,l} a(k,l)]$, where $\max_{k,l} a(k,l)$ is the max value of $a(k,l)$, is divided into 1000 bins, and we choose the bin b according to the condition

$$\sum_{i=i}^{b} h(1000-i) \geq p \sum_{i=0}^{999} h(i). \qquad (6.30)$$

The bin number in the histogram $b \in [0, 999]$ and p is a predefined parameter that allows fine-tuning of the exact location of b in the histogram. In Bardeli (2009), this parameter is set to $p = 0.96$. Finally, the threshold θ_a is obtained as

$$\theta_a = \frac{b}{1000} \max_{k,l} a(k,l). \qquad (6.31)$$

In Figure 6.12, we present the manually generated thick black lines, which illustrate the application of this method for the spectrogram shown in Figure 6.1. The

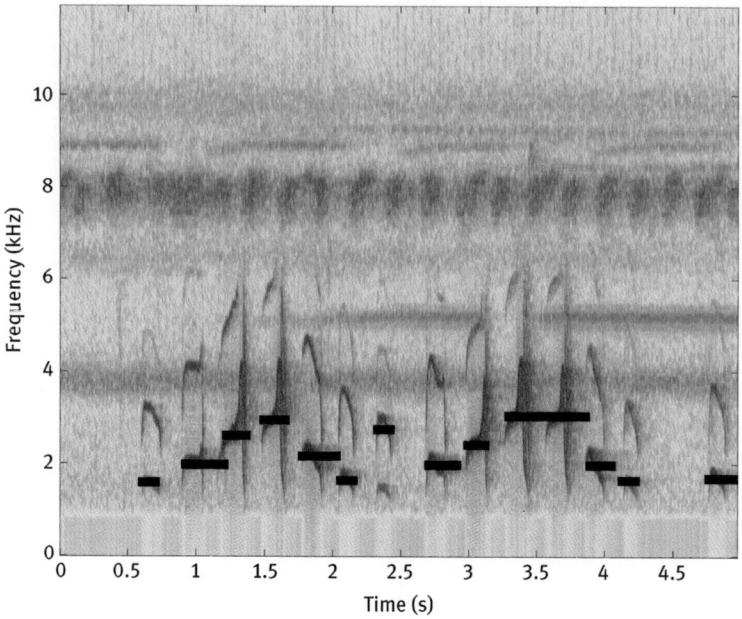

Figure 6.12: Points of interest are located along the thick black line that corresponds to pixels with non-zero attention value. The manually generated image is based on the spectrogram shown in Figure 6.1.

result will be the estimation of numerous points of interest along the thick black line that corresponds to pixels with non-zero attention value $a(k,l)$.

For each point of interest, a 4-D feature vector (i,k,l,c) is assigned. The tuples consists of the index ith of the current audio signal, the indices k and l of the frequency band and the time segment, which are determined depending on the current position in the spectrogram, and the class c. The class c was defined as a 12-bit number, where each bit describes some aspect of the geometric shape of the 2-D DFT computed for a neighbourhood of 16×16 pixels with centre the selected point of interest (Bardeli 2009). The value of each bit in the feature class c is set zero or one after comparison of the row and column values of the 2-D Fourier transform with heuristically chosen threshold.

6.6 Bag-of-instances audio descriptors

As explained earlier, the 1-D audio parameterization methods discussed in Chapter 5 and the hybrid 2-D/1-D method discussed in Section 6.4 compute audio descriptors that enclose information about all frequency bands within a certain predefined frequency range. This holds true for all cepstral coefficient-like audio parameterization schemes, such as the LFCC, MFCC, GFCC, and so on. The fact that these descriptors enclose information about all frequency bands complicates the subsequent modelling and decision-making, especially in situations where multiple sound events co-occur in time, even if they only partially overlap in frequency range with the target sound events. In fact, these complications are significant even when sound events of different species do not overlap in frequency range but concur in time, particularly in the species identification task where we aim to identify a number of species, so narrowing the frequency range is not always applicable.

This specificity of audio descriptors computed through 1-D audio parameterization methods is considered a significant disadvantage in application set-ups that require the implementation of the multi-label species identification task (cf. Section 3.3), where we aim to identify all species that are acoustically active in an audio chunk. This is particularly important in biodiversity monitoring applications, which make use of soundscape recordings captured with omnidirectional microphones, and aim to identify multiple sound events of multiple species within each chunk of audio. This application scenario is referred to as multi-instance multi-label classification (Zhou and Zhang 2007a; Briggs et al. 2012).

Here, we outline an audio parameterization method proposed in Briggs et al. (2012), which makes use of 2-D processing of the audio spectrogram considered as a greyscale image. This method is applicable for multi-instance multi-label classification and has relaxed requirements to the preparation of training and validation libraries (cf. Chapter 4). In particular, the audio recordings in the training library only need to be tagged for the list of species that are acoustically active in the specific audio file, without any need of a syllable-level annotation of time boundaries, or of selective manual editing for the elimination of prominent interferences.

In brief, Briggs et al. presented an audio parameterization method, which estimates a set of audio descriptors, referred to as *bag-of-instances*, for each chunk of audio. Each audio chunk is considered a *bag*, which combines the individual fixed-length audio descriptors for all *instances*. Thus, the bag-of-instances represents the combination of all relevant *instances* contained within the spectrogram computed for an audio chunk. Each *instance* is a feature vector computed for a single automatically identified 2-D object, referred to as a *blob*. The 2-D objects may correspond to a birdcall, a syllable in a call sequence, or to a short passage in a song. This method separates all non-overlapping 2-D objects in the time–frequency plane and then computes audio descriptors for each individual object. The audio descriptors for a given instance usually portray the temporal evolution of a certain spectral peak between its start and end time.

The computation of audio descriptors is summarized in Figure 6.13. Specifically, the audio pre-processing steps, including the audio pre-filtering, audio segmentation to short quasi-stationary segments, and the spectrogram computation, follow the signal processing flow presented in Section 6.1. In the work of Briggs et al. (2012), audio pre-filtering was not used, and it does not seem mandatory when recordings with high SNR are processed. This is because the method implements an iterative sub-band noise reduction in the time–frequency domain.

When the method is applied on real-world recordings, where low-frequency interferences due to wind blow in the microphone and mechanical vibrations, noise from rain and other sources can have significant amplitude, it makes sense to use high-pass pre-filtering, especially when the low-frequency sub-bands are not of interest. Such a pre-filtering would suppress the DC component and the low-frequency interferences and would help to reduce the dynamic range of DFT coefficients computed via eq. (6.1). The last facilitates the subsequent 2-D noise reduction and the audio parameterization methods, which operate on the audio spectrogram. This also helps to save processing time and energy and is especially important when the pre-processing and computation of descriptors are performed on mobile devices with fixed-point CPU and smart sensors.[3] However, once pre-filtering is applied to reduce noise in the low-frequency sub-bands,

---

[3] The Iot concept heavily relies on such mobile devices and smart sensors as these are seen as the means for information acquisition.

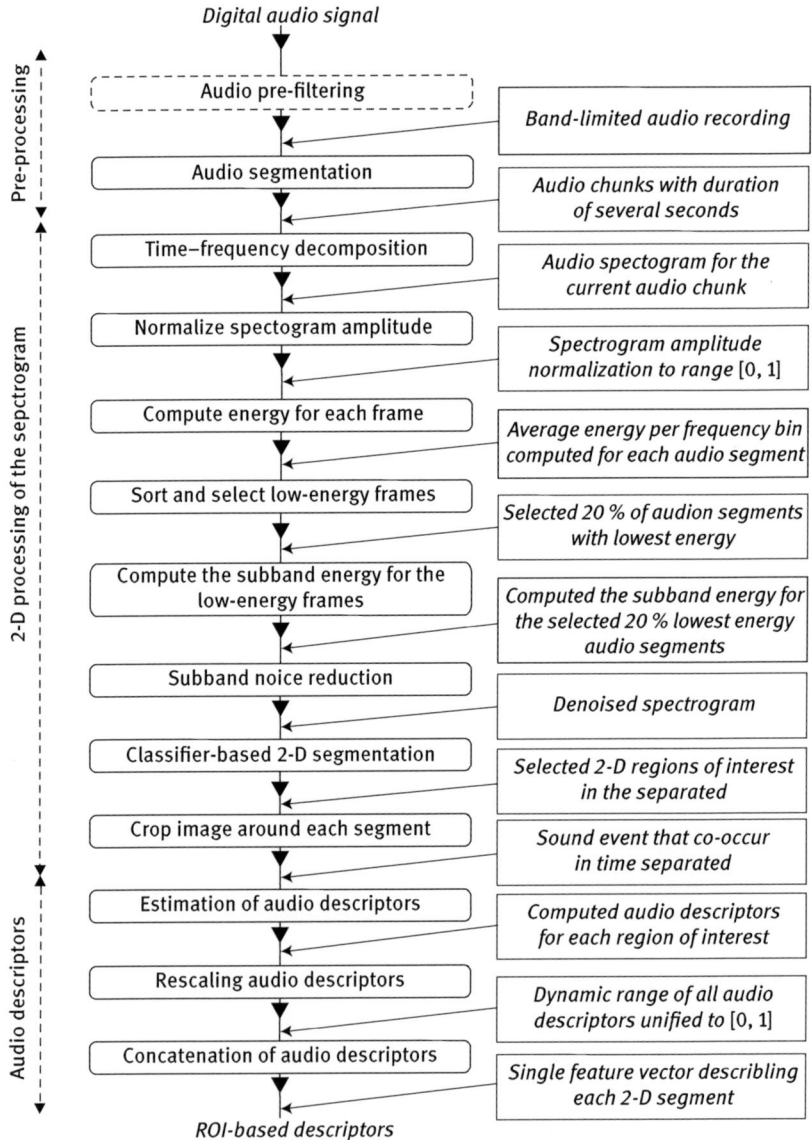

Figure 6.13: The overall block diagram of the ROI-based audio parameterization method.

these sub-bands should be excluded from processing with the 2-D noise reduction method used in Briggs et al. (2012) because any residual impulsive noise will be amplified again.

Here, we adhere to the original exposition in Briggs et al. (2012) except that we used audio pre-filtering with a high-pass filter (cut-off 50 Hz) to eliminate the

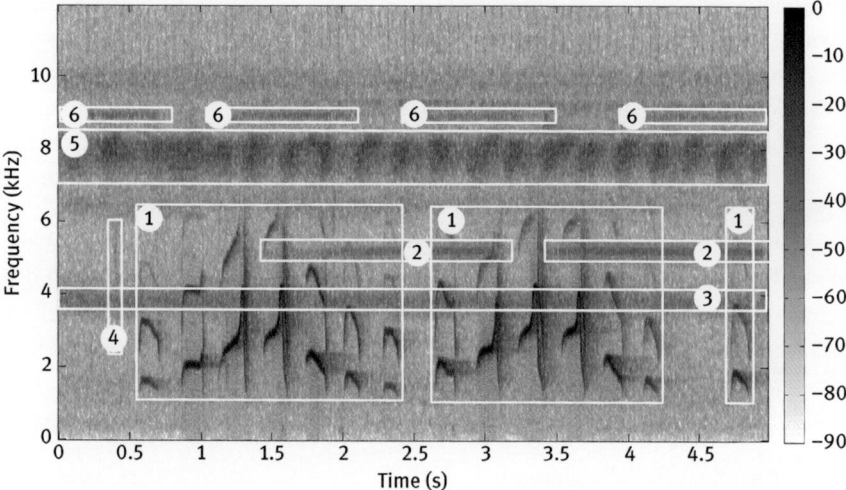

Figure 6.14: The audio spectrogram obtained from the original audio signal, after pre-filtering with a high-pass filter (cut-off 50 Hz).

DC offset and lowest frequency bands. After the audio pre-processing and computation of the spectrogram (Figure 6.14), all magnitudes in the spectrogram are normalized to dynamic range [0, 1]. Similarly to all previous spectrogram plots, Figure 6.14 is presented in dB scale.

The 2-D processing of the spectrogram starts with an iterative sub-band noise reduction, which operates on the normalized by amplitude spectrogram. The noise reduction algorithm, which is also referred to as *whitening filter*, implements iterative re-estimation of the sub-band noise profile over time based on the low-energy segments. The noise reduction algorithm consists of two iterations, each implementing the following three steps:
1. Selection of audio segments that contain only noise
2. Estimation of the noise profile in each frequency sub-band
3. Compensation of noise based on the noise profile estimated in each sub-band

In order to select audio segments that contain only noise, we first compute the average amplitude across all frequency bins k in the spectrogram for each audio segment l

$$E(l) = \frac{1}{N} \sum_{k=0}^{N-1} |S(k,l)|, \; 0 \leq l \leq T, \tag{6.32}$$

and then sort in ascending order all $E(l)$. A subset E_{20}, containing 20 % of the audio segments with lowest energy, is kept for the estimation of the noise profile:

$$P(k) = \sqrt{\varepsilon + \sum_{l \in E_{20}} |S(k,l)|^2}, 0 \le k \le N-1, \quad (6.33)$$

where $\varepsilon = 10^{-10}$ is a small constant helping to prevent $P(k) = 0$ for all k.

In order to facilitate the detection of ROI and the subsequent image segmentation, each frequency sub-band of the spectrogram is compensated based on the estimated noise profile $P(k)$:

$$S_2(k,l) = \frac{\sqrt{|S(k,l)|}}{P(k)U(l)}, 0 \le k \le N-1, 0 \le l \le T \quad (6.34)$$

where the units column $U(l)$ is a vector of T elements. The product $P(k) U(l)$ is a matrix with the same size as $|S(k,l)|$.

The result after the first iteration of this whitening filter is shown in Figure 6.15. As presented in the figure, the sound emissions from insects (e. g. sound events tagged with /2/, /3/, /5/, and /6/) have been suppressed significantly, without heavy damage of the sound events of the target species /1/, which have mostly vertical structure, that is, wideband spread across the frequency axis and narrow presence over time. However, due to the selective suppression of frequency bands, there is some irregularity in the spectrogram and some increase of

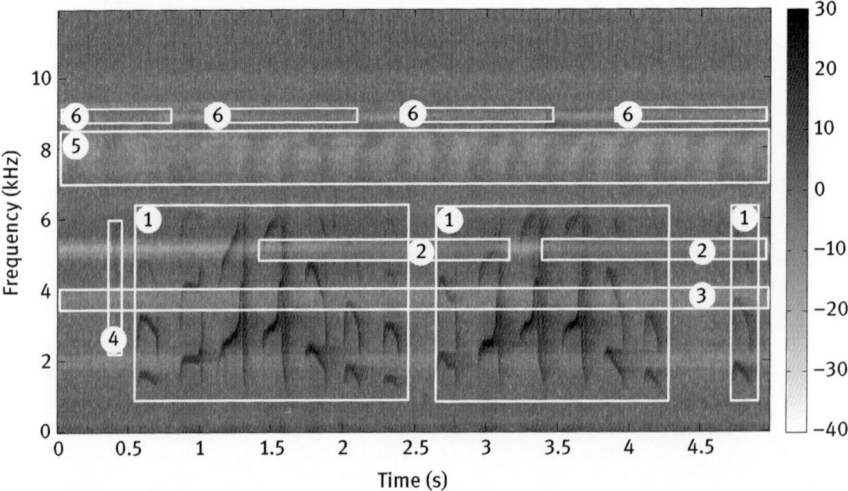

Figure 6.15: The spectrogram after the first iteration of noise reduction algorithm of the spectrogram shown in Figure 6.13.

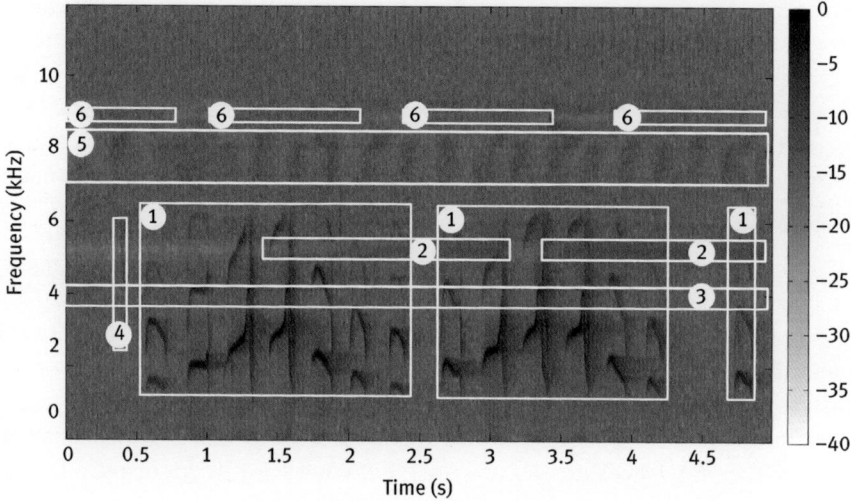

Figure 6.16: The spectrogram after the second iteration of the noise reduction algorithm.

noise amplitude in the lowest frequency band below 50 Hz. The last is due to the operation division (6.33) and to the very low values estimated for $P(k)$ in eq. (6.32) in the frequency sub-bands that have been pre-filtered.

After the second iteration of the whitening filter, all major interferences (tagged with /2/–/6/) were suppressed significantly and the spectrogram[4] in Figure 6.16 looks cleaner and more homogeneous when compared to the spectrogram in Figure 6.14.

When we compare the original spectrogram (Figure 6.14) and the cleaned one (Figure 6.16), we realize that the noise suppression algorithm of Briggs et al. (2012) is quite efficient in dealing with long-duration interferences, such as the sound events tagged with /2/, /3/, /5/, and /6/. These tags correspond to sound emissions of insects and exhibit little variability in frequency and amplitude over time, whereas the noise suppression algorithm preserves the fast-changing spectral peaks of the target species /1/ and the interference from unknown origin tagged with /4/.

In general, such noise suppression method would be quite efficient for the suppression of long songs of frogs and toads, but also of sound emissions of other

---

4 The spectrogram shown in Figure 6.16 was normalized to dynamic range [−40, 0] dB to facilitate the comparison with the original spectrogram (Figure 6.14) and the spectrogram obtained after the first iteration (Figure 6.15) of the whitening filter. The genuine dynamic range of magnitudes for the spectrogram in Figure 6.16 is [−35, −15] dB.

animals that are characterized with long-duration monotonous sound sequences and low variability over time.

Briggs et al. (2012) employed two slightly different configurations of this noise suppression method, depending on the purpose of the subsequent steps – audio segmentation and audio parameterization. Specifically, for the purpose of audio segmentation Briggs et al. applied only the first iteration of the whitening filter using a different estimate of the noise profile:

$$P(k) = \frac{1}{\operatorname{card}(E_{20})} \sum_{l \in E_{20}} |S(k,l)|, 0 \le k \le N-1, \tag{6.35}$$

and without applying a square root on the spectrogram in eq. (6.34), which becomes

$$S_2(k,l) = \frac{|S(k,l)|}{P(k)U(l)}, 0 \le k \le N-1, 0 \le l \le T. \tag{6.36}$$

The designation $\operatorname{card}(E_{20})$ stands for the cardinality of the subset E_{20}, which contains the values $E(l)$ for the selected low-energy audio segments, that is, the number of elements in E_{20}.

These changes do not alter the performance of the whitening filter; however, these allow that the cleaned spectrogram preserves the large dynamic range of spectral magnitudes (Figure 6.17).

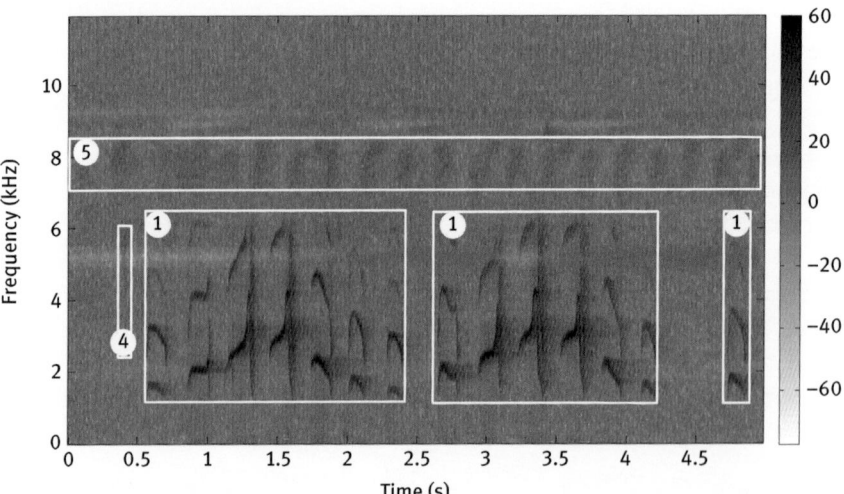

Figure 6.17: Spectrogram obtained with the second variant of the whitening filter (6.36) applied on the spectrogram shown in Figure 6.14.

A side effect of the application of this noise suppression method, in both its configurations discussed here, is that it might introduce spectral gaps, that is, artefacts in the frequency sub-bands where the energy drops fast and then recovers to average levels again. This happens in the portions corresponding to silence in the frequency sub-bands, where there are intermittent sound events from insects, some frogs, and other species that emit long repetitive sounds with little-to-no variability in frequency and amplitude over time. However, this inconveniency is minor and the method is prised with good noise suppression capability for this particular type of interferences.

Once the spectrogram is processed with the whitening filter, the image needs to be segmented to ROI. The detection of potentially important ROI can be implemented with a supervised classification method, which is trained with manually annotated audio spectrograms in order to learn the difference between target sounds and noise. In Briggs et al. (2012), the image segmentation is implemented with a random forest classifier with 40 trees. However, the classifier architecture and settings depend on the resolution of the audio spectrogram and on the specifics of the sound events of interest. The outcome of such segmentation is blobs, which confine the ROI (Figure 6.18). Here, we consider the favourable case, where the interference tagged with /5/ were suppressed sufficiently by the whitening filter, and therefore, none of the image irregularities in that frequency range was selected by the segmentation algorithm. The short-duration interference tagged with /4/ is represented with a single blob.

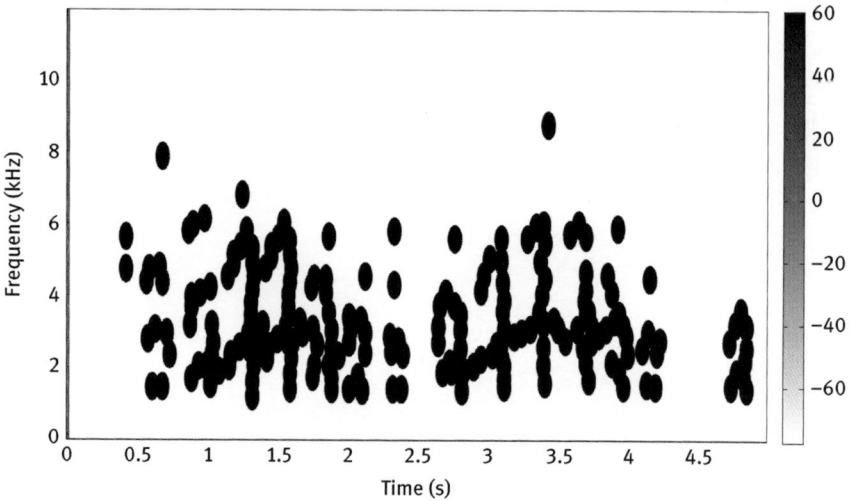

Figure 6.18: Manually generated time-frequency mask for the spectrogram in Figure 6.17.

Once all 2-D objects in the spectrogram are identified, the spectrogram is cropped so that each object becomes an independent image. This allows the spatial separation of sound events, which co-occur in time but do not overlap in frequency. When two non-overlapping in frequency domain sound events overlap in time, we first use the mask of the first event to select all audio segments under the mask and then zero all frequency components outside the mask. Then the procedure is repeated for all sound events that co-occur in the same time interval.

Audio descriptors are computed for each 2-D object obtained to this end, except for small objects that have few pixels. In particular, Briggs et al. (2012) computed three types of descriptors:
1. *Mask descriptors*
2. *Profile statistics*
3. *Histogram of gradients*

The eight mask descriptors (*min-frequency, max-frequency, bandwidth, duration, area, perimeter, non-compactness*, and *rectangularity*) carry information about the size and shape of the mask, and not the audio segment below it.

The profile statistics descriptors carry information about the statistical properties of the time and frequency profiles of the sound event segment under the mask. These are represented by 16 descriptors, including 14 that are based on the normalized time and frequency profile (the *normalized time and frequency probability densities*, their *Gini index, mean, variance, skewness, kurtosis,* and *maxima*) plus the *mean* and *standard deviation* of the spectrogram under the mask region.

The shape and texture of each segment is characterized by the 16-dimensional histogram of gradients-based descriptors, which are based on the cropped spectrogram under the mask and the mask. The cropped spectrogram is convolved with a Gaussian kernel and the partial gradients over time and frequency are computed. Only pixels with sufficient energy, which exceeds a predefined threshold, contribute to the 16-bin histogram. Bins are evenly spaced over the range of angles $[0, 2\pi]$. The 16 histograms of gradient-based descriptors are the normalized counts in each bin of the histogram.

Finally, the values of the 40 audio descriptors computed to this end are normalized to dynamic range $[0, 1]$ and are next concatenated to form a single feature vector for each 2-D object. In the work of Briggs et al. (2012), the normalized feature vector is next used in multi-instance multi-label classification algorithms (Zhou and Zhang 2007a, 2007b; Zhang and Wang 2009; Zhang 2010) that aim to identify all sound events within an audio chunk.

Concluding remarks

The 2-D audio segmentation and audio parameterization methods discussed in Chapter 6 illustrate some of the most distinctive and influential concepts, which offer appealing ideas for further research on 2-D noise reduction and 2-D audio feature extraction. They all make use of image processing techniques applied on the spectrogram considered as an image and offer opportunities for deployment in real-world applications where *offline* processing of soundscape recordings is required.

Hardware and software limitations, deficiency of energy resources, and time restrictions impose various practical considerations on the manner signal preprocessing and audio parameterizations are implemented. Due to these practical considerations, the implementation of 2-D audio parameterization methods requires that long audio recordings be partitioned to a number of non-overlapping chunks with duration in the range of few seconds to a minute. Next, each chunk is processed independently of the others, and due to the inherent variability of environmental noise, usually re-estimation of the noise profile and adjusting of the noise suppression settings is required. The buffering of audio in the range of several seconds to a minute is the main source of delay, and the additional time for signal and image processing sums up on the top of that. Thus, the cumulative delay could be of over a minute with respect to current time.

The advantages of 2-D audio segmentation and 2-D audio parameterization, discussed in Chapter 6, are well supported by the results of several recent technology evaluation campaigns. Among these are the NIPS4B 2013 Multi-label Bird Species Classification,[5] LifeCLEF Bird Identification Task 2014[6] (Goeau et al. 2014), and LifeCLEF Bird Identification Task 2015[7] (Joly et al. 2015), where the winning systems (Lasseck 2013, 2014, 2015) made use of 2-D spectrogram-based methods for audio segmentation and parameterization. These were combined with 1-D audio parameterization followed by extending the 1-D audio descriptors with various statistical derivatives (Eyben et al. 2013) and also made use of publicly available metadata provided by the competition organizers. The main strengths of the bird species identification systems developed by Lasseck (2013, 2014, 2015) are due to the careful implementation of 2-D methods for noise suppression, the reliable audio segmentation, and the large number of audio features used, including

---

[5] Neural Information Processing Scaled for Bioacoustics: NIPS4B, http://sabiod.univ-tln.fr/nips4b/
[6] LifeCLEF-2014 Bird task, http://www.imageclef.org/2014/lifeclef/bird
[7] LifeCLEF-2015 Bird task, http://www.imageclef.org/2015/lifeclef/bird

the use of species-specific subsets of audio features. In these technology evaluations, it has been demonstrated that the 2-D audio features outperform the large set of 1-D audio features and the metadata-based classification. However, a well-conceived fusion of information from various information streams was shown beneficial for improving the overall recognition accuracy.

7 Audio recognizers

Introduction

The mathematical description of the probabilistic modelling was formulated by Thomas Bayes (1763), who laid the foundations of the statistical decision-making theory. In brief, Bayes' theory is grounded on two assumptions: (i) that the decision problem can be specified in probabilistic terms and (ii) that all relevant probabilities are known or can be derived from the observed data and/or some prior knowledge about the problem. These two assumptions are in common use in the contemporary philosophy of statistical classification. Bayes' theory has laid the groundwork for the elaboration of a great number of classification methods.

7.1 Overview of classification methods

The implementation of the various species detection, identification, recognition, and the sound events recognition tasks discussed in Chapter 3 builds on 1-D or 2-D audio parameterization (cf. Chapters 5 and 6) and some classification methods. Here, we briefly outline the taxonomy of classification methods commonly employed in these tasks. Most of these classification methods were firstly employed in speech and audio processing-related tasks. Nevertheless, for simplicity of exposition, in the current section we refer only to the species recognition task, though all arguments are applicable to the other tasks discussed in Chapter 3.

In Figure 7.1, we show some members of the two major categories of classification methods: *discriminative* and *non-discriminative*, as well as a few of their combinations. A comprehensive presentation on the principles and specifics of generative and discriminative machine learning methods is available in Jebara (2004). In the following, we briefly point out some aspects and emphasize the main differences among these.

The *discriminative methods* are trained to minimize the classification error on a set of training data. Thus, they *only* aim to estimate the boundary between different categories and are insensitive to the intraclass variability of data within each category. Among the discriminative machine learning methods are
- linear discriminant analysis (Fisher 1936),
- polynomial classifier (Specht 1966, 1967),
- time-delay neural networks (Lang and Hinton 1988),
- recurrent neural networks (RNNs) (Jordan 1986; Pearlmutter 1989; Elman 1990),

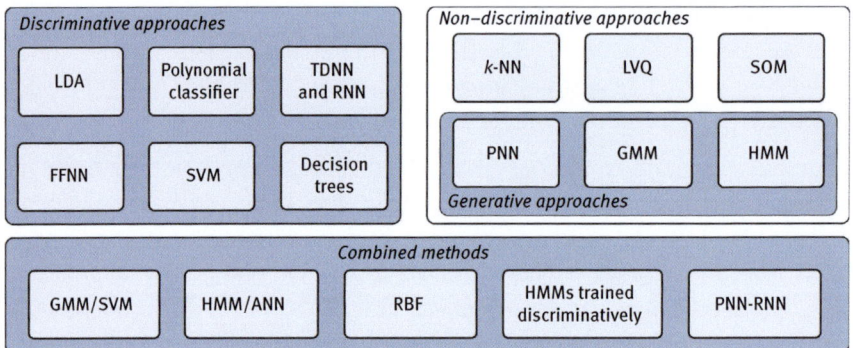

Figure 7.1: Taxonomy of some widely used classification methods.

- multi-layer perceptron (MLP) neural networks (Rosenblatt 1958),
- support vector machines (SVMs) (Vapnik 1995), and
- decision trees (Breiman 2001; Geurts et al. 2006; Lasseck 2013).

The *non-discriminative* methods do not aim at minimizing the classification error via direct adjustment of the decision boundary, and instead group the data by category in a certain way. A major subdivision of the non-discriminative methods is referred to as *generative*. As their name implies, the *generative classification methods* aim to create a model of the underlying distribution of each category, relying on the available training data sets.

Among the commonly used generative methods are the
- probabilistic neural network (PNN) (Specht 1988, 1990), which combines Parzen-like probability density function estimators with Bayes' strategy for decision rules;
- Gaussian mixture models (GMMs) (Pearson 1894; Hansen 1982); and the
- hidden Markov models (HMMs) (Baum and Petrie 1966; Baum 1972).

HMMs are capable of modelling the temporal correlations among events in sequences of observations. GMM (as a single-state HMM) is not sensitive to the temporal order in series of observations, and thus the model creation is computationally less demanding. Likewise, the original PNN (Specht 1988) also treats the subsequent observations independently one from another, and thus is indifferent to any temporal correlations in data sequences. However, the most distinctive trait of all generative approaches is that they treat the data of each category independently from those of the other categories.

In general, the generative methods possess more adjustable parameters when compared to the discriminative ones. The reason for having more adjustable parameters lies in the pursuit to create appropriate model of the intraclass distribution of observations in each category. The last is not explicitly required in the various species and sound event recognition tasks; however, it might help in dealing with outliers, improving stability in mismatched conditions, and reducing the misclassification rates. This is because in applications of the species and sound events recognition technology, which operate with soundscape recordings and field audio recordings, the generative methods tend to be more resistant to noise. This is mainly because (i) they are able to create comprehensive model(s) of the acoustic events of interest and (ii) often create an explicit model of the acoustic environment and the major sources of noise and interferences in order to compensate for their effects. This feature turns out to be a precious virtue in noisy real-world environments.

In addition to the *generative* classification methods, the group of the *non-discriminative* methods includes also other members that cannot be labelled as generative because they do not model data densities. Among these are the k-nearest neighbour (k-NN) classifiers (Cover and Hart 1967), self-organizing maps (SOMs) (Kohonen 1981, 1982; Somervuo and Kohonen 1999), learning vector quantization (LVQ) (Kohonen 1995; Somervuo and Kohonen 1999), and so on. Strictly speaking, these are not likelihood estimators, and therefore, they do not belong to the generative approach. The *non-discriminative* methods are characterized with lesser discriminative power and are rarely used alone in contemporary species recognition applications. However, these were reported beneficial in the intermediate steps of data processing, and thus are typically used in combination with other machine learning techniques.

Both *discriminative* and *generative* approaches have limitations, and none of them provides a perfect classification results in all practical applications. These limitations are due to the assumptions on which these classification methods rely and are well explained by the *no-free-lunch-theorem* (Wolpert 1996; Wolpert and Macready 1997).

Furthermore, in an attempt to overcome some of these limitations and shortcomings a large number of hybrid classification approaches were created. As shown in Figure 7.1, in conjunction with the above-mentioned two major approaches (*discriminative* and *generative*), there exists a densely populated third category, the members of which simultaneously possess properties characteristic of the discriminative and generative methods. The group of the combined methods is comprised of

1. hybrid classifiers, which merge generative model with a discriminative classifier; and
2. discriminatively trained generative classifiers, whose parameters are adjusted by optimizing a certain discriminative objective function.

Among these hybrid methods are the
- Radial basis function (RBF) (Powell 1987; Broomhead and Lowe 1988), which unites the generative GMMs (in fact, unimodal Gaussian densities) with the discriminative MLP.
- GMM-LR/SVM (Bengio and Mariethoz 2001), which builds generative GMM into the framework of discriminative SVM.
- HMM/MLP (Bourlard and Morgan 1994) combine the temporal capabilities of the HMM with the discriminative capacity of the MLP. Two other tandems, namely HMM/RNN (Neto et al. 1995) and the input/output HMM (Bengio and Frasconi 1996), exploit the same division of functions as in the HMM/MLP approach.
- Discriminatively trained HMM (Setlur et al. 1996) belong to the second subdivision of the combined methods, namely the discriminatively trained generative classifiers. In the discriminatively trained HMM, the maximum likelihood criterion of the traditional HMM is replaced by the maximum mutual information criterion.
- PNN-RNN hybrids, among which are the locally recurrent probabilistic neural networks (LRPNNs) (Ganchev et al. 2003; Ganchev 2009a, 2009b), Generalized locally recurrent probabilistic neural networks (GLRPNNs) (Ganchev et al. 2007) and partially connected locally recurrent probabilistic neural networks (PCLRPNNs) (Ganchev et al. 2008). These combine the advantages of RNN and PNNs. The main difference among the LRPNN, GLRPNN, and PCLRPNN is in the recurrent layer linkage and in the presence/absence of memory at the input of recurrent layer neurons.

All PNN-RNN hybrids elaborate on the PNN architecture by embedding a RNN as an additional hidden layer between the pattern layer and the decision layer of the original PNN. In this way, the PNN-RNN hybrids combine the positive traits of the generative and the discriminative approaches that are expressed in generating robust models, which are supported by improved discriminative capabilities. However, the most important aspect of the LRPNN and GLRPNN hybrids is their capability to exploit temporal correlations in sequences of observations. In the case of species recognition, the PNN-RNN architectures are capable of exploiting the interdependence among audio descriptors computed for successive audio segments, which contributes significantly to improved performance and robustness on field recordings. Further details on the applicability of these methods to speech and audio-related tasks are available in Ganchev (2005) and Potamitis and Ganchev (2008).

The great variety of discriminative, non-discriminative, generative, and combined classification methods offers opportunity for selecting an appropriate classifier for each specific task; however, this is not a trivial problem. A comparative evaluation of all machine learning methods on a certain task is not straightforward due to the huge amount of effort required for training of models and fine-tuning of the classifiers. Furthermore, many research groups publish on proprietary data sets, which are not publicly available, use dissimilar experimental set-ups, and thus the empirical results reported are not directly comparable among different studies.

Besides the assortment of classifiers, data sets, and experimental set-ups, the choice of audio descriptors and the variety of post-processing schemes contribute significantly to divergence of the reported empirical results. As the correct way for a direct comparison among methods is to carry out performance evaluation on a common data set and experimental protocol, in the past years the bioacoustic research community established a number of technology evaluation campaigns. In this regard, an encouraging progress was made within the Kaggle[1] challenges on computational bioacoustic tasks and the annual species recognition challenges LifeCLEF organized in the recent years (cf. Chapter 9). Among these were

- NIPS-2013 Multilabel bird species classification,[2]
- MLSP-2013 Bird classification challenge,[3]
- ICML-2013 Bird challenge,[4]
- LifeCLEF-2014 Bird task,[5]
- LifeCLEF-2015 Bird task,[6]
- EADM-2015 Challenge,[7] and
- LifeCLEF-2016 Bird task.[8]

Other similar challenges, which are not mentioned here, were organized on tasks related to the acoustic recognition of whale and insect sound emissions.

However, since the participation in such technology evaluation campaigns is voluntary, only few of the aforementioned machine learning approaches were evaluated there. Another difficulty is that the settings of the algorithms for which the results were obtained during these evaluations sometimes remain confidential to the participants. Besides the general ranking of participants, not

---

[1] The Kaggle website, https://www.kaggle.com/competitions
[2] The NIPS-2013 Multi-label bird species classification task, https://www.kaggle.com/c/multilabel-bird-species-classification-nips2013
[3] The MLSP-2013 Bird classification challenge, https://www.kaggle.com/c/mlsp-2013-birds
[4] The ICML-2013 Bird challenge, https://www.kaggle.com/c/the-icml-2013-bird-challenge
[5] The LifeCLEF-2014 Bird task, http://www.imageclef.org/2014/lifeclef/bird
[6] The LifeCLEF-2015 Bird task, http://www.imageclef.org/lifeclef/2015/bird
[7] The EADM-2015 Challenge, http://sabiod.univ-tln.fr/eadm/#challenge
[8] The LifeCLEF-2016 Bird task, http://www.imageclef.org/lifeclef/2016/bird

all methods and results are published with comprehensiveness, which would facilitate exact reproduction of the experimental results. Nevertheless, the public technology evaluations contribute to a better understanding of the advantages and limitations of the different audio parameterization and machine learning methods under the specific application-oriented set-up and experimental protocol.

Currently, GMM, HMM, SVM, and decision tree classifiers are considered state of the art in the species recognition technology. However, according to the no-free-lunch-theorem, other machine learning methods might lead to a better performance, depending on the specifics of the particular species or sound event recognition task and the available acoustic libraries. Specifically, during the past decade, GMM-based modelling (Reynolds 1995; Lee et al. 2013; Henríquez et al. 2014; Ganchev et al. 2015) and HMM-based modelling (Baum 1972; Trifa et al. 2008a, 2008b; Bardeli et al. 2010; Chu and Blumstein 2011; Potamitis et al. 2014; Oliveira et al. 2015; Ventura et al. 2015) were the most frequently used methods for acoustic animal identification. However, in the past few years, the SVM classifiers that were firstly used for speaker identification in Schmidt and Gish (1996) were mastered on the species recognition tasks as well. Alternative methods, such as the hybrid GMM/SVM (Bengio and Mariethoz 2001), also have potential to achieve highly competitive results.

Lately, decision trees and more specifically the random forest decision trees are the preferred choice for multi-label classification tasks, especially when 2-D audio segmentation and 2-D audio parameterization is involved (Breiman 2001; Neal et al. 2011; Briggs et al. 2012; Lasseck 2015). This is because the multi-label random decision trees make feasible the multi-label classification to be performed in a computationally efficient manner (Zhang et al. 2010), when compared to the other multi-instance multi-label (MIML) classifiers, and in some occasions with better accuracy. The transition from the traditional single-instance single-label supervised object classification methods, where the classifier names one category per audio chunk (as if the chunk contains sounds of a single dominant species or a single sound event), towards the MIML modelling paradigm, where the classifier has to identify all categories (species or sound events) present in an audio chunk, brought a major advance to the computational bioacoustic methods.

In brief, the MIML paradigm was developed in the past decade for the needs of image processing technology and then was adapted to the needs of 2-D audio segmentation and 2-D audio parameterization. These recent advances are especially beneficial in the automated recognition of species in soundscape recordings. A nice overview of MIML classification methods is available in Gibaja and Ventura (2014) and Zhang and Zhou (2014), and a tutorial on multi-label learning in Gibaja and Ventura (2010). Software implementations of various multi-label classification algorithms and multi-instance methods, as well as examples and demos on various data sets, are publicly available through the LAMDA group,[9,10] *scikit-learn*,[11] MEKA[12], and so on (cf. Chapter 9). Many of the MIML learning methods were already demonstrated appropriate to the multi-label species identification task and the applicability of other still needs to be evaluated.

Finally, ensemble learning methods (Okun et al. 2011; Zhang and Ma 2012; Zhou 2012), which combine the predictions of several machine learning models, have been used as well. The ensemble learning methods offer more flexibility as they can model more complex functions and combine the advantages of various individual classifiers.

In the following sections, we outline the overall system architecture used in the typical computational bioacoustic tasks, which illustrates the manner in which some audio parameterization and machine learning approaches are used and combined. Specifically, in Sections 7.2–7.4 we discuss automated detectors and recognizers, implemented for the needs of the one-species detection and the species identification tasks, which make use of state-of-the-art single-label classification methods. Finally, in Section 7.5 we discuss the fundamentals of MIML classification methods. Advances in MIML methods will enable the deployment of new technological tools and services, and therefore, these possess the potential to provide improved technological support to biodiversity-related studies.

7.2 One-species detectors

The GMM – universal background model (GMM-UBM) and the HMM-based modelling approaches are considered the current state-of-the-art in the one-species detection (cf. Section 3.1) and the one-category recognition (cf. Section 3.4) tasks. This is mainly due to their ultimate advantage when compared to discriminative classifiers (cf. Section 7.1), namely they explicitly model the intraclass distribution of audio descriptors for both the target species and the acoustic background. All this provides an effective mechanism for coping well with outliers and for compensation of environmental noise, which is of essential importance when

---

9 Prof. Z.-H. Zhou's Publications, http://cs.nju.edu.cn/zhouzh/zhouzh.files/publication/publication_toc.htm
10 The Institute of Machine Learning and Data Mining (LAMDA), Nanjing University, China, http://lamda.nju.edu.cn/MainPage.ashx
11 The *scikit-learn* machine learning in Python, http://scikit-learn.org/stable/
12 MEKA: A Multi-label Extension to WEKA, http://meka.sourceforge.net/

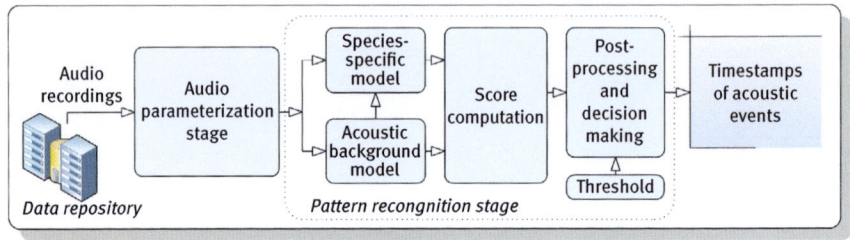

Figure 7.2: Overall block diagram of a GMM-UBM-based species-specific recognizer.

working with soundscape recordings. This is because usually the soundscape recordings are quite noisy and the co-occurrence of multiple sound events is common. Therefore, when special corrective measures are not taken, such as noise suppression and segmentation of sound events, high error rates are observed. The main disadvantage of the GMM-UBM- and HMM-based methods is their dependence on manually tagged, or manually annotated, training libraries for the target species and a large acoustic background library that is free of sounds of the target species (cf. Chapter 4). When such acoustic libraries are not available, or when these are not representative for the acoustic condition in which the recognizer operates, the misclassification rates increase considerably.

In brief, the GMM-UBM likelihood test approach was originally introduced by Reynolds et al. (2000) on the speaker verification task. Later on this approach was successfully applied to other audio processing tasks, including on species identification and species recognition tasks (Graciarena et al. 2010; Ganchev et al. 2015). Summarizing the exposition in Ganchev et al., we briefly outline the architecture of an automated species-specific recognizer based on the classical GMM-UBM approach. The overall block diagram of a species-specific acoustic recognizer is shown in Figure 7.2.

Specifically, given a sequence of audio feature vectors $\mathbf{X}$, computed for a sequence of audio frames and the hypothesized animal species S, the task of automated species detection consists in determining whether $\mathbf{X}$ originates from the target species S. From this perspective, the task consists in testing the following two hypotheses:

H_0: $\mathbf{X}$ originates from the hypothesized species S,
H_1: $\mathbf{X}$ does *not* originate from the hypothesized species S.

Given a sequence of audio feature vectors $\mathbf{X}$, and given a mechanism for the reliable estimation of likelihoods of the hypotheses H_0 and H_1, we can apply the optimum test (7.1) and decide between these two hypotheses:

7.2 One-species detectors — 145

Likelihood ratio test	Comparison with a threshold and decision making
$\dfrac{p(\mathbf{X}\|H_0)}{p(\mathbf{X}\|H_1)}$	$\geq \theta$ accept H_0, (Decision: **X** originates from the species S), $< \theta$ reject H_0, (Decision: **X** does *not* originate from the species S).

(7.1)

Here, $p(\mathbf{X}|H_0)$ and $p(\mathbf{X}|H_1)$ are the probability density functions for the hypotheses H_0 and H_1 evaluated for the sequence of audio feature vectors **X**, which is computed for the observed audio segment. The decision threshold θ for accepting or rejecting H_0 needs to be adjusted based on a representative validation library.

In practice, H_0 is represented with the mathematical model λ_{hyp}, which characterizes the hypothesized species S in the audio feature space. Likewise, $\lambda_{\overline{\text{hyp}}}$ represents the alternative hypothesis H_1. Then, the likelihood ratio (7.1) can be rewritten as

$$\frac{p(\mathbf{X}|\lambda_{\text{hyp}})}{p(\mathbf{X}|\lambda_{\overline{\text{hyp}}})}. \tag{7.2}$$

Furthermore, most often we make use of the logarithm of eq. (7.2) in order to reduce the dynamic range of values. Such an arrangement gives the representation of the log-likelihood ratio (LLR) as

$$\Lambda(\mathbf{X}) = \log\left[p(\mathbf{X}|\lambda_{\text{hyp}})\right] - \log\left[p(\mathbf{X}|\lambda_{\overline{\text{hyp}}})\right]. \tag{7.3}$$

The model λ_{hyp} for H_0 is well defined and can be estimated using the training library for the target species S. However, the model $\lambda_{\overline{\text{hyp}}}$ for H_1 is not specified well as it has to represent all possible alternatives to the hypothesized target species.

Given a collection of audio recordings from a large number of species that are representative of the community of sound-emitting species observed in the habitat, a single model $\lambda_{\text{UBM}} \sim \lambda_{\overline{\text{hyp}}}$ is built to represent the alternative hypothesis. It is also possible to use multiple background models tailored to specific sets of species or habitats; however, the use of a single-background model has advantages in terms of computational efficiency.

An important step in the implementation of the likelihood ratio test is the selection of the actual likelihood function $p(\bar{\mathbf{x}}_t|\lambda)$. The choice of this function heavily depends on the audio features being used as well as on the specifics of the application. The mixture density used for the likelihood function is defined as follows:

$$p(\bar{\mathbf{x}}_t | \lambda) = \sum_{i=1}^{M} w_i p_i(\bar{\mathbf{x}}_t). \tag{7.4}$$

In this way, the mixture density is seen as a weighted linear combination of M unimodal Gaussian densities $p_i(\bar{\mathbf{x}}_t)$, each parameterized by a $D \times 1$ mean vector $\mathbf{\mu}_i$ and a $D \times D$ covariance matrix Σ_i. The mixture weights w_i satisfy the constraint $\sum_{i=1}^{M} w_i = 1$. Here $p_i(\bar{\mathbf{x}}_t)$ are defined as

$$p_i(\bar{\mathbf{x}}_t) = \frac{1}{(2\pi)^{D/2} |\Sigma_i|^{1/2}} \exp\left\{-\frac{1}{2}(\bar{\mathbf{x}}_t - \mathbf{\mu}_i)' \Sigma_i^{-1} (\bar{\mathbf{x}}_t - \mathbf{\mu}_i)\right\} \tag{7.5}$$

While the general form of the model supports full covariance matrices, typically we assume that the audio descriptors are decorrelated, and therefore we can use simply the diagonal covariance matrices and thus reduce the number of adjustable parameters. Collectively, the parameters of the density model are denoted as $\lambda = \{w_i, \mathbf{\mu}_i, \mathbf{\sigma}_i\}$, where $i = 1, 2, \ldots, M$.

Given a set of audio feature vectors computed from the training library, maximum likelihood model parameters can be estimated using the iterative expectation-maximization (EM) algorithm (Dempster et al. 1977; Bishop 2006). The EM algorithm iteratively refines the GMM parameters to increase monotonically the likelihood of the estimated model for the observed feature vectors. Under the assumption of independence among subsequent feature vectors, the log-likelihood of a model λ for a sequence of T audio feature vectors is computed as

$$L(\lambda) = \frac{1}{T} \sum_{t=1}^{T} \log [p(\bar{\mathbf{x}}_t | \lambda)]. \tag{7.6}$$

Often, eq. (7.6) is computed for a sliding sequence of $T \in [3, 10]$ vectors as this allows achieving good time resolution in the identification of start and end timestamps of each acoustic event.

Aiming at computational efficiency, it is convenient to use a single-background model to represent $p(\mathbf{X} | \lambda_{\overline{\text{hyp}}})$. Using a GMM as the likelihood function, the background model λ_{UBM} is typically implemented as a large GMM trained to represent uniformly the distribution of audio descriptors for the specific acoustic environment. Specifically, the acoustic background library is usually selected in such a way as to reflect the expected acoustic environment encountered during the operation of the detector as well as all significant sources of interference.

The model λ_{hyp} for the target species S is obtained as adapted GMM, which is derived by adapting the parameters of the background model λ_{UBM} via maximum a posteriori (MAP) estimation (Reynolds et al. 2000) using the training library for S. Since the data set representing the target species is usually quite small, the MAP adaption only updates the mean vectors of the model. Because the target species model λ_{hyp} is created by updating the parameters in the acoustic background model λ_{UBM}, there is a close association between the corresponding mixture components of the two models. The last facilitates the discriminative capability of the species-specific detector.

The decorrelation stage in the 1-D audio parameterization (cf. Chapter 5, eq. (5.8)) permits the use of diagonal covariance mixture density GMMs that have fewer free variables in their covariance matrices to estimate, and therefore, are better trained to represent the target classes when the training libraries are quite small (Reynolds et al. 2000). In the case when we rely on 2-D audio descriptors (cf. Chapter 6), which are not necessarily decorrelated, we have to estimate the full covariance matrix density GMMs. The last offers the opportunity for more accurate modelling of the underlying distribution of these audio descriptors; however, this assumes the availability of a larger training data set and incurs larger computational demand.

Computing the LLR (7.3) for a sequence of audio feature vectors **X** requires estimation of the likelihood twice: first for the target species model, and second for the acoustic background model. This is repeated for each audio feature vector in **X**, and therefore is computationally expensive for models with a large number of mixtures. However, the fact that the hypothesized species model λ_{hyp} is adapted from the background model λ_{UBM} allows for using fast scoring methods (Reynolds et al. 2000; Saeidi et al. 2009).

Since the components of the adapted GMM retain a correspondence with the mixtures of the acoustic background model, the feature vectors close to a particular mixture in the background model will also be close to the corresponding mixture in the target model. For that reason, instead of scoring all mixtures of the acoustic background model and the target model for each audio feature vector, we can determine the top-C scoring mixtures in the background model and compute the background model likelihood using only these top-C mixtures (Reynolds et al. 2000). Typically, we score the audio feature vector against only the corresponding components in the adapted target species model to evaluate the likelihood. For instance, in the case of background model with $M = 1024$ mixture components and $C = 10$ scoring mixtures, this would require only $M + C$ computations per feature vector, which is much less when compared with the $2M$ computations for the case of exhaustive likelihood ratio evaluation.

In summary, each sequence of audio feature vectors **X** (e. g. 1-D audio features computed as explained in Section 5.2 and post-processed as in Section 5.3, or 2-D audio descriptors computed as in Chapter 6) is subject to the log-likelihood test

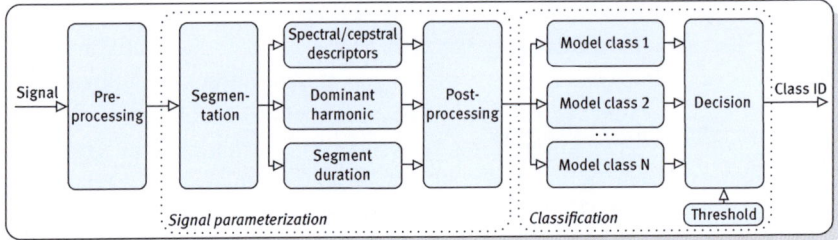

Figure 7.3: Multi-class classifications with a single model per category.

(7.3) applied on a frame-by-frame basis. The scores obtained in this manner are next compared with the decision threshold θ, which results in a series of binary decisions, one per frame:
- "Yes" – the audio segment corresponds to sound event related to the target species.
- "No" – it belongs to a sound of different origin.

These decisions are often convoled with a rectangular window in order to correct sporadic misclassifications due to bursts of interfering audio signals. The post-processed decisions are then transformed to timestamps indicating the onset and end of each sound event in relative or absolute time units.

7.3 Single-label classification with a single model per category

In statistical machine learning approaches, such as the generative methods PNNs (Specht 1988), GMMs (Reynolds 1995), and HMMs (Baum 1972) it is typical to build an individual model for each category of interest. Alternatively, in the discriminative and combined approaches, such as the SVM (Vapnik 1995), GMM/SVM (Bengio and Mariethoz 2001), and so on, we typically create a separate binary classifier for each category of interest.

In Figure 7.3, we show the general block diagram of a typical species identification system, regardless of the choice of a particular classification method. This diagram illustrates the main signal processing and pattern recognition steps typical to a single-label, multi-class, open-set, classifier with a single model per target category.

Here, the notion single label refers to the fact that the classifier assigns a single tag to each chunk of the signal or to the entire input. Multi-class refers to the fact that the system is capable of assigning multiple different tags, that is, there are more than one species (or sound events) of interest, which could be distinguished one from another. Open set refers to the fact that the input might be different from the categories for which there are prebuilt models. It is typical that each species, or each sound event, of interest is modelled independently from the others. The multi-class tasks are frequently reduced to a number of two-class problems, where instead of category-specific models we make use of category-specific classifiers.

In brief, after acquisition the audio signal is subject to *pre-processing*, which may implement removing of the DC component, adaptive gain control, band limiting of the signal, compensation of spectral slope, noise removal, and so on. The pre-processed signal is then subject to *signal parameterization*, which consists of the following three main steps (cf. Chapters 5 and 6):

- *Audio segmentation*, which splits the audio signal on small stationary portions, usually with uniform length, referred to as audio segments or audio frames.
- Computation of *signal descriptors* (aka audio parameterization or audio features extraction), which includes parameters computed in the temporal domain, frequency domain, cepstral domain, wavelet domain, supra segmental parameters, and so on.
- *Post-processing* of the signal descriptors for removing their mean value, dynamic range normalization, lessening the effects of environmental noise, variations due to the equipment type, set-up, settings, and other circumstances, which cannot be controlled during the audio acquisition process. The post-processing of audio features vectors facilitates the convergence of model training and the classification process.

The sequence of post-processed audio features vectors is next used in the *classification* stage. The audio feature vectors can consist of audio descriptors computed on frame or on chunk level. During classification with generative methods, the sequence of audio feature vectors, computed for an audio chunk, is compared against all pre-built models. A Bayes-optimal decision is made based on the computed likelihoods the input to belong to each specific category. In discriminative methods, decision is usually made based on the computed score with respect to a decision boundary or compared to a threshold. In both generative and discriminative methods, the final decision might integrate comparison with a decision threshold, for example, in open-set classification tasks, however, this is not a requirement.

The essentials here are that each unlabelled input sequence has to be compared to all N pre-built models or to be processed by N two-class classifiers. In order to reduce computational demands, certain methods rely on pre-selection of the models to which comparison will be made. For instance, the pre-selection can be based on k-NN or some sorting index function (Saeidi et al. 2009).

A common trait of all single-label classification methods, regardless of whether they are implemented with discriminative, generative, or combined classifier, is that each input audio chunk is assumed to correspond to a single ground true tag or composed of a number of (non-overlapping) single-tag pieces. Each piece is assumed to contain a single sound event, sound emission of a single species, or that sounds of one species dominate above the noise floor. This assumption is rarely met for soundscape recordings made with a single-channel omnidirectional microphone, which is the common case in automated biodiversity studies. This is a major drawback, which limits the applicability of all single-label methods. This deficiency is partially alleviated when field recordings are made with a parabolic directional microphone or when microphone arrays are used during the audio acquisition.

The use of directional microphones is common in traditional bioacoustic studies as it allows to reduce significantly interferences and acoustic background noise, and to consistently capture predominantly the sounds originating from the fauna individual of interest. However, since individuals are moving, typically a human presence is required to point the directional microphone in the correct direction. Therefore, directed parabolic microphones are mostly applicable in limited-scale, limited-duration, and human expert-attended audio recording campaigns. The involvement of humans is the reason for significant logistical expenses and bias towards short-term monitoring in more accessible habitats and favourable weather conditions. The presence of humans increases the risks of disturbance, and together with the typical limited duration of observation campaigns may not allow capturing the whole richness of behaviours. Due to the dependence on human labour, scalability to large area studies is not achievable.

On the other hand, the use of multi-microphone arrays in studies relying on automated recorders (Trifa et al. 2008b; Mennill and Vehrencamp 2008; Collier et al. 2010) allows broad area monitoring and capturing of sounds from moving individuals, as well as their localization and tracking, while providing a good suppression of background interference from other sources. However, these benefits are efficient mostly at short distances, which depend on the size and geometry of the microphone array set-up. Furthermore, the use of microphone arrays increases significantly the required audio storage capacity because good practices require that the original audio recordings for all channels be kept for archival and future study needs. A comprehensive overview on the applicability of microphone arrays and arrays of wireless acoustic sensors to recognition, localization, and tracking of individuals and other tasks of computational bioacoustics is available in Blumstein et al. (2011).

7.4 Single-label classification with multi-classifier schemes

In most real-world applications of machine learning technology, including those considered in computational bioacoustic tasks, no classifier can provide perfect recognition results. Therefore, in order to achieve satisfactory recognition accuracy, often we rely on collaborative multi-classifier schemes. The multi-classifier schemes permit models based on different assumptions and methodology to complement each other, and in collaboration to achieve better recognition accuracy than each individual classifier can obtain alone (Hansen and Salamon 1990; Ho et al. 1994; Jordan and Jacobs 1994; Lam and Suen 1995; Breiman 1996; Dietterich 1998; Hinton et al. 1998; Kittler et al. 1998; Alkoot and Kittler 1999; Dietterich 2000; Kittler and Alkoot 2003; Bishop 2006).

As we did in Section 7.3, here we consider that an individual expert is created for each target category (Figure 7.4). The signal parameterization stage, including the audio pre-processing, segmentation, feature extraction, and feature post-processing steps, is similar to the signal processing steps outlined in Section 7.3. However, in the case of multi-classifier schemes, each expert has a multifaceted structure with multiple models or multiple classifiers tuned to detect the same specific target class. The number of classifiers engaged in each individual category-specific expert and/or the audio features they use do not need to be identical among the different categories. In fact, quite frequently different classifiers, developed for a particular category of sounds, make use of dissimilar feature vectors. This increases the chances that each classifier would provide complementary information, which is advantageous for increasing the overall recognition accuracy of the category-specific expert.

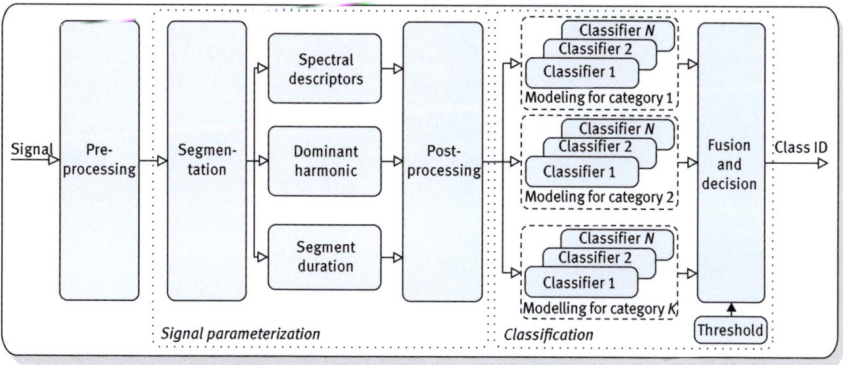

Figure 7.4: Multi-class classifier with multiple models per category.

Making an optimal decision, based on the combination of individual classifiers outputs in a compound category-specific expert, is not a trivial problem. Various hierarchical and ensemble schemes for combining classifiers were studied in the literature (Xu et al. 1992; Kittler et al. 1998; Alkoot and Kittler 1999; Dietterich 2000). Many of these schemes are based on achieving balance in the bias–variance reduction trade-off, and others are based on plurality voting, majority voting, or other more elaborative voting methods. Here, we briefly mention two simple cases:
1. linear score-level fusion of parametric and non-parametric generative classifiers (GMM and PNN), and
2. plurality voting of heterogeneous classifiers (e. g. SVM and GMM).

These are easy to implement and demonstrate reliable performance in various sound classification tasks. However, non-linear fusion and classifier-based fusion schemes would usually outperform these.

In addition to the final decision, the generative classifiers can also provide some confidence measure, which indicates the degree of confidence, this decision to be correct, or not. The last enables efficient fusion schemes to be put into effect, regardless whether these operate on the classifiers' output scores or decisions.

In particular, a score-level fusion of PNN and GMM classifiers has been reported beneficial on the speaker recognition problem (Ganchev et al. 2002) and on the acoustic insect identification problem (Ganchev and Potamitis 2007). The gain of classification accuracy after fusion is mainly due to the different ways that the PNN and GMM estimate the underlying distribution of feature vectors for the target classes. Specifically, during training, these approaches aim to approximate the underlying distribution of some post-processed audio descriptors, computed for each target category of sounds, which eventually results in a certain statistical model. During the operational stage, these models are exploited to compute the degree of resemblance between each model and an unlabelled input signal. The outcome of each comparison is a score (usually expressed in terms of post-probabilities or likelihoods), which is next exploited to make the final decision. The generative approaches are capable of estimating the confidence in the final decision they made, and thus offer additional information and transparency that contribute towards an improved understanding of the classification results.

In brief, a linear combination scheme for score-level fusion is presented as

$$P_k = \sum_{j=1}^{J} \alpha_{jk} P_{jk}, \qquad (7.7)$$

where P_k is the resulting score obtained through fusion of the post-probabilities of the individual classifiers trained for class k, α_j is the weight coefficient for the jth classifier for class k, and P_{jk} is the post-probability for the jth classifier for class k. For the specific case of two generative classifiers, namely the PNN and GMM, the score-level fusion can be written as

$$P_k = \alpha_k P_{\text{GMM},k} + (1 - \alpha_k) P_{\text{PNN},k}, \tag{7.8}$$

where $P_{\text{GMM},k}$ and $P_{\text{PNN},k}$ are the post-probabilities for the GMM and PNN classifiers for a specific class k. In the general case, the weights α_k can have class-specific values. Often an equal significance of the all classifiers is assumed, and $\alpha_k = 1/J$ is used for all target classes. Here, J is the number of fused classifiers. When generative classifiers are involved in score-level fusion schemes, the Bayesian decision rule is not applied to the outputs of the individual classifiers, but instead to the fused scores P_k, computed via eq. (7.8).

Based on such a fusion scheme, Ganchev and Potamitis (2007) developed class-specific experts for acoustic recognition of insects, built around a combination of PNN and GMM classifiers. Such a collaborative scheme takes advantage of prior knowledge about the following specifics of these classifiers: the PNN offers robustness and good generalization capabilities in deficiency of training data (i. e. tiny training library), while the GMM is capable of building fine-tuned models when abundant training data is available. Thus, given a sufficient amount of training data, GMMs are capable of modelling the underlying distribution in a more accurate manner than the PNN. However, since GMMs have multiple adjustable parameters they require more computational efforts during training and larger amounts of training data, which is not always available.

The voting collaboration schemes provide a simple way for improving the overall accuracy when heterogeneous classifiers need to be combined, and especially when one (or more) of these classifiers does not provide a confidence measure for their decision. Specifically, in plurality voting, each classifier has one non-weighted vote per given input observation, and this vote can support any of the target categories. The winning class is the one which accumulates most of the votes, regardless of whether majority is achieved or not. Thus, in multi-class classification problems the final decisions per input sample are often made without majority. The last is often pointed out as a disadvantage of the plurality method. However, from another point of view, this mechanism secures decision for every input observation, and no inputs are wasted due to lack of majority among the classifiers.

7.5 MIML classification schemes

As discussed in Section 3.3, the multi-label species identification task aims to identify all species that are acoustically active in a certain audio chunk. The final goal is to list the species discerned in that audio chunk, that is, only to identify the set of species tags. The recognition of time–frequency boundaries of each sound event, discerned in the specific audio chunk, might not be required.

Variations in the amplitude, duration, and structure of sound events among species cause difficulties in the species and sound events recognition tasks. However, the greatest challenge among all is due to the overlap among sound events in the time–frequency space. Since in the multi-label species identification task we assume that an audio chunk may contain sounds of more than one species, we regard the MIML classification paradigm as the apparent direct approach for implementing the simultaneous identification of multiple instances. An alternative indirect approach to address this problem would be to iteratively scan throughout the audio chunk with species-specific detectors, identify the loudest sound event, and erase it from the audio chunk. Next, we continue searching for the next loudest sound event, find it, erase it, and continue until the desired number of categories is attained. Alternatively, the iterations can continue until the noise floor level is reached and all sound events that are discernible are already enumerated. In the following discussion, we focus on the direct MIML classification approach.

The MIML classification paradigm has been introduced firstly for the needs of image processing applications and afterwards a plethora of classifiers based on this idea were developed for various other applications (Zhou and Zhang 2007a). The advantages of MIML classifiers in terms of classification accuracy on computational bioacoustic tasks were demonstrated in the literature (e. g. Briggs et al. 2012; Briggs 2013; Briggs et al. 2013a; Lasseck 2013, 2015) and their performance was comparatively evaluated in several technology evaluation campaigns focused on the multi-label species identification.

In order to take advantage of the MIML classification paradigm in the multi-label species identification task, we consider that an entire audio chunk of several seconds is a single object. This object might contain one or more sound events emitted from a single individual or several individuals that belong to one species or different species. Here, each sound event we refer to as an *instance*, and therefore an audio chunk could normally contain multiple *instances*. Collectively all instances within a single audio chunk are referred to as *bag of instances*, and the class labels are attributed to each bag and not to each instance. Each label is attributed to the bag only if there is at least one instance belonging to that class.

Well, if we can establish a mechanism that makes possible a group of sound events in an audio chunk to be associated with the corresponding species *label(s)*,

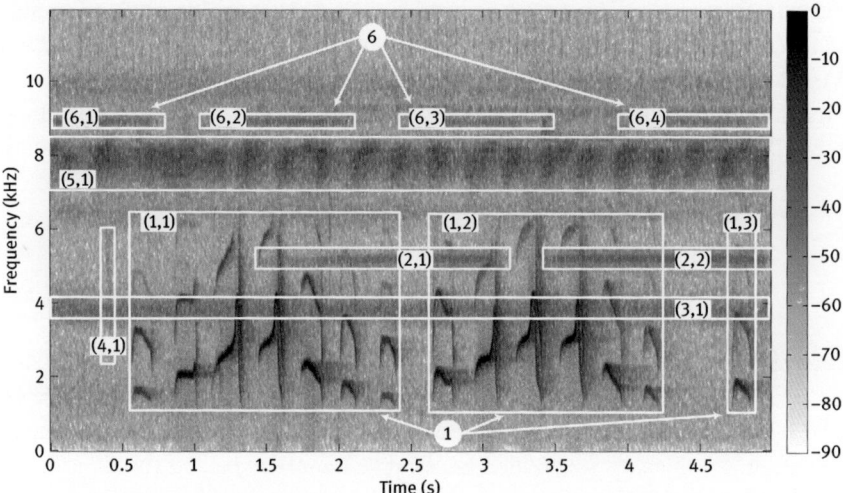

Figure 7.5: The audio spectrogram manually annotated as a multi-instance multi-label object. The pairs of numbers for each sound event stand for the species tag and the instance index.

we open the opportunity to end up with a list of all species that are acoustically discernible in that specific audio chunk.

The MIML classification paradigm usually takes advantage of 2-D audio segmentation and 2-D audio parameterization methods (cf. Chapter 6) in combination with certain machine learning methods, which are based on the MIML classification paradigm.

In order to illustrate the multi-label species identification problem, let us consider the audio spectrogram shown in Figure 7.5, which was computed for a 5-s excerpt of a real-world recording.[13] As seen in the spectrogram, this chunk of audio contains three sound events of the dominant species (Amazonian Streaked Antwren, *Myrmotherula multostriata*) tagged with /1,1/, /1,2/, and /1,3/. The first digit in each pair stands for the species tag /1/ and the second for the instance index. As shown in the spectrogram, several sound events of other species were captured along with the sound emissions of the dominant species. They appear

---

13 A 5-s excerpt from recording MYRMUL09.mp3 (offset 13 s from the start). The dominant species tagged with /1,1/, /1,2/, and /1,3/ in Figure 7.5 is the Amazonian Streaked Antwren (*Myrmotherula multostriata*), recorded by Jeremy Minns, 2003-04-10, 15:58h, Brazil, Pousada Thaimaçu. Rio São Benedito. Source: Xeno-Canto.
  Available at http://www.xeno-canto.org/sounds/uploaded/DGVLLRYDXS/MYRMUL09.mp3

with a lower energy and yet high above the noise floor to be visually and acoustically discernible. Here, these sound events are provisionally tagged with labels /2,1/, /2,2/, /3,1/, /4,1/, /5,1/, /6,1/, /6,2/, /6,3/, /6,4/.

A faultless multi-label classification algorithm applied on the audio chunk shown in Figure 7.5 would output a list of six species. Of course, the correct species identification depends on the proper segmentation of all acoustic events, which is one of the major challenges to address (cf. Chapter 6). Therefore, once again we need to emphasize that achieving an accurate segmentation is not a trivial task as the sound events may possibly overlap in both time and frequency range. For instance, this is the case with the sound events tagged with /2,1/, /2,2/, and /3,1/, which overlap in time and frequency range with other sound events. Such an overlap may cause some sound events, or important parts of them, to remain entirely unobserved, or to be split in isolated fragments. Both types of segmentation errors may possibly cause misdetection or misclassification, regardless of the type of audio descriptors and the classification approach used afterwards – this is simply because improper segmentation might cause loss of important details which cannot be compensated on a later stage of sound event identification.

The conceptual architecture of a multi-instance multi-species identification system is shown in Figure 7.6.

This species identification system consists of three main stages: 1-D signal processing, 2-D image processing, and a multi-instance multi-species classifier. The 1-D signal processing stage takes as an input audio signal and implements audio pre-processing, time–frequency transform, and computation of the audio spectrogram. The pre-processing step typically includes automated gain control, downsampling, reduction of the frequency range, and so on. The short-time

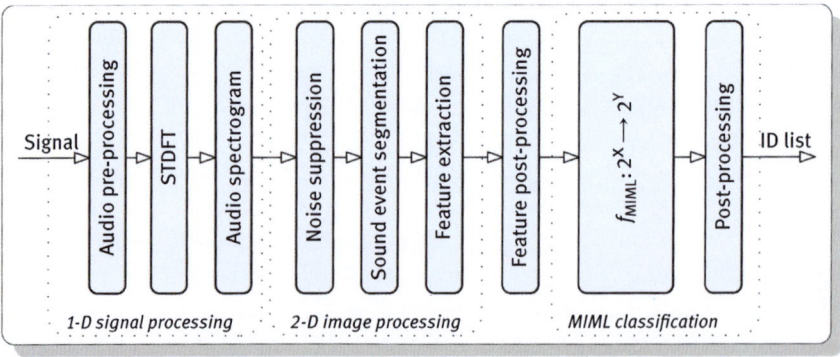

Figure 7.6: System architecture of multi-species identification by means of 2D-audio processing of the spectrogram and multi-instance multi-label classifier.

discrete Fourier transform (STDFT splits the pre-processed audio signal to overlapping segments, with duration of a few milliseconds to few tens of milliseconds, and computes the discrete Fourier transform for each segment. This process effectively converts the 1-D signal to a matrix of coefficients, where each row is the Fourier spectrum of the corresponding audio segment. The selected segment duration and windowing function type define the time–frequency resolution of the STDFT,[14] and therefore, one needs to carefully select settings which allow to obtain an acceptable trade-off between temporal and frequency resolution of the analysis. Subsequently, the audio spectrogram is computed by squaring the Fourier coefficients or, in some cases, by taking their absolute value.

Once the audio spectrogram is computed, it is considered a greyscale image, which is fed as input to the 2-D image processing stage. The 2-D image processing leads to some appropriate set of descriptors that well capture the time–frequency properties of the sound events of interest and suppress unwanted sources of variability. This usually involves some image processing-based method for noise suppression, which removes the background of the image and thus emphasizes the sound events with compact localization of energy in the time–frequency space. Segmentation of the image to regions, each containing an entire sound event, is afterwards applied in order to select the ROI. Each ROI is next subject to a parameterization, which aims to compute a set of sound event descriptors. Some concepts for the computation of sound event descriptors via image processing-based techniques were discussed in Chapter 6.

The set of descriptors needs to be representative and relevant to the categories or interest. In addition, these descriptors also have to suppress any undesired sources of variability of audio signal due to interferences from the environment. Because most often the descriptors obtained to this end do not respect these requirements, a post-processing is applied. A post-processing of descriptors might include
- standardization (zero mean, unit standard deviation),
- dynamic range scaling to [0, 1],
- mapping to a lower-dimensional space in order to reduce some undesired variability,
- concatenation of the raw descriptors with their temporal derivatives and/or higher-level features obtained by computing some higher-order statistics on the original low-level features, and so on.

---

14 The Gabor limit stipulates that within a single STDFT decomposition one cannot achieve high resolution simultaneously in the time and in the frequency domain.

Finally, the post-processed feature vectors are fed to the MIML classifier, which follows the multi-label classification paradigm introduced by Zhou and Zhang (2007a), and is adapted to the needs of automated species identification as in Briggs et al. (2012). Specifically, in the MIML machine learning approach each ROI in the audio spectrogram is a multidimensional set of feature vectors $X_j = \{\mathbf{x}_{j,1}, \mathbf{x}_{j,2}, \ldots, \mathbf{x}_{j,n}\}$ that is a *bag of instances* assumed to embody a set of objects. Each object could represent individual call notes, syllables, or other continuous portions of sound, and each bag is considered a subset of the multidimensional feature space. In the common case, these objects may take any label of a set of known labels of interest $Y_j \subseteq \{\mathbf{y}_{j,1}, \mathbf{y}_{j,2}, \ldots, \mathbf{y}_{j,n}\}$, where the number of labels in the subset Y_j may vary. Therefore, the classifier aims to map a set of bags into a set of labels making use of some MIML function $f_{MIML} : 2^X \to 2^Y$, where X represents the entire multidimensional feature space and Y represents the set of labels for the classes of interest.

The MIML learning and classification problem ($f_{MIML} : 2^X \to 2^Y$) can be solved in two ways – either through the multi-instance learning or through the multi-label learning approaches (Zhou and Zhang 2007a; Zhou et al. 2012) by transforming the MIML problem into a

- Multi-instance task ($f_{MI} : 2^X \to Y$): the multi-instance task can be further transformed into the traditional supervised single-instance single-label problem ($f_{SISL} : X \to Y$) (Xu and Frank 2004).
- Multi-label task ($f_{ML} : Z \to 2^Y$), where certain mapping function φ transforms the multi-instance bags to single-instance input ($\varphi : 2^X \to Z$). The multi-label task can be further transformed into the supervised single-instance single-label problem ($f_{SISL} : Z \to Y$) (Zhou and Zhang 2007b).

The MIMLBOOST and MIMLSVM methods (Zhou and Zhang 2007b) are examples of classifiers implementing these two approaches, respectively. Besides, various other classifiers were adapted to the MIML classification problem by transforming the multi-instance classification to a number of single-instance problems (Huang et al. 2014). Among these are the MIML k-NN (Zhang 2010), a hidden conditional random world model (Zha et a. 2008), and a generative model for MIML learning (Yang et al. 2009).

The most significant disadvantage of the MIML approaches is the high computational complexity, and therefore, MIML classifiers are applicable only on small size to moderate data sets. Operation on large data sets is complicated due to the prohibitive computational demand. In order to deal with this difficulty, Huang et al. (2014) developed a fast MIML approach (MIMLfast) which operates in a lower-dimensional subspace. This is a significant step towards making the MIML approaches scalable, and therefore, applicable to large data sets, and to computational bioacoustic task that require dealing with big data. Here, big data refers to continuous recording of soundscapes in the range of months or full annual cycles.

Concluding remarks

To this end, we illustrated the conceptual design of few species or sound events recognition tasks and outlined some classification schemes, which provide valuable technological support towards scalable bioacoustic studies. However, the automated classification and recognition of sound events are not the only conceivable way to support scalable biodiversity assessment and biodiversity monitoring studies.

In this regard, Glotin et al. (2013a) discussed sparse coding for soundscape analysis that aims to obtain optimal information encoding. Such encoding is considered optimal in the sense that it would enclose enough information to reconstruct the signal near regions of high data density, while the reconstruction of signal in regions of low data density is not essential. The sparse coding concept was considered in a forest soundscapes analysis scenario. Among the main conclusions of this study was that the use of a dictionary of time–frequency atoms facilitates the analysis of forest sound organization without the need of prior classification of individual sound events. Certainly, the sparse information encoding approach does not exclude developments, which implement classification of sparsely coded sound events.

Eldridge et al. (2016) brought a further advancement of sparse coding and source separation with the aim to access and summarize ecologically meaningful sound objects. This opens possibilities to investigate the composition of soundscapes in terms of dynamic interactions between spectra-temporal patterns of species vocalizations. More important here is that background sounds are no longer seen as noise that has to be eliminated. Instead, these are perceived as important components, which are needed for the proper interpretation of the interaction between sounds created by wind, rain, and other natural phenomena, human activity, and animal communities. While their study does not provide definite answers on how such statistical analysis will be used in biodiversity studies, it points out a number of valid research questions that need to be addressed in order to understand the relation between properties *complexity* and *decomposability* and the status of an acoustic community.

We do not see these and other parallel studies, which support the perspective of ecoacoustics and this aim to advance the holistic assessments of soundscapes, as antagonists to computational bioacoustic technology discussed in this book. Instead, we deem that these provide relevant tools and services that are complementary and vital for the implementation of scalable biodiversity assessment and biodiversity monitoring studies. Furthermore, these provide information at different level of abstraction (and thus disregard different information sources) and are nice examples on how technology could support the efforts for biodiversity preservation in various ways.

8 Application examples

Introduction

As discussed in Chapter 1, biodiversity is essential to preserving the capacity of earth to support complex life. Furthermore, alongside economic, cultural, and ethical aspects, preserving the genetic richness of Earth's ecosystem is also important for future technological advances of humankind. Throughout the entire history of our civilization, we have studied the living nature in order to gain knowledge, inspiration, or insights in possible ways to improve our technology. Biodiversity is of particular importance for technology advancement because flora and fauna species comprise a vast repository of ideas and information, which directly contribute to the diversity of concepts and developments in the areas of genetics, inheritance and evolution, pharmacology, robotics, artificial life, artificial intelligence, and so on.

Unfortunately, at present the extinction rate of species is estimated to be faster than the speed of new species emergence, as well as their discovery and study by science. According to raw estimations of experts,[1] the number of unidentified species for certain orders and families is larger than the number of those already studied. Due to fragmentation and loss of habitats, pollution, and other human-induced pressure, many of these unidentified species will be extinct before we have the opportunity to study them. The consequences of such a loss are unpredictable.

Unsurprisingly, public and private entities already provide sensible support to the growing efforts for slowing down the worrisome trend of rapid biodiversity loss.[2] In that connection, biodiversity monitoring, assessment, and conservation initiatives are perceived as the way to go, and therefore relevant actions receive significant public attention. These initiatives aim to enhance our understanding about biodiversity-related problems and to investigate opportunities for mitigation of potential threats due to biodiversity loss. In brief, biodiversity preservation is already recognized

1. essential to the health and sanity of our civilization,
2. of great importance to economy and culture, and

---

[1] The LifeWatch project, http://www.lifewatch.eu/
[2] Here, we focus only on aspects related to acoustic monitoring of terrestrial animal species and its importance to biodiversity assessment in support of focused conservation actions. Due to space limitation, we do not discuss the acoustic monitoring of aquatic species, the use of other (non-acoustic) biodiversity monitoring methods, and the harmful effects due to human-induced pollution of environment with substances and energy.

DOI 10.1515/9781614516316-008

3. as a precondition for the very survival of human species and complex life on earth.

The enhanced public awareness about all these motivates the urgent need of actions on a global scale. The last advised the establishment of international agencies and funding bodies in support of local, regional, and global biodiversity-related research and conservation actions. For example, in Europe among these are EC LIFE+,[3] Biodiversa,[4] EU Structural and Investment Funds,[5] and INTERREG Europe,[6] in the USA primary support comes through the National Science Foundation[7] (NSF), in Brazil through the National Council for Scientific and Technological Development[8] (CNPq), CAPES Foundation,[9] and so on. These and other funding sources across the world provide a number of financial instruments in support of large-scale international and intercontinental initiatives[10,11] that coordinate diverse activities aiming at slowing down the global trend of rapid biodiversity loss.

Furthermore, the benefits of interdisciplinary activities and large-scale biodiversity preservation undertakings are well illustrated through the achievements of a number of recent research and technology development projects, which took advantage of such an international collaboration. Specifically, in the following sections we overview a few characteristic projects with significant social impact. These made a confident use of computational bioacoustic methods and tools in order to address a range of challenging real-world problems. Furthermore, these projects are interesting because they made the difference with respect to previous research, despite the hardware and software limitations at the time of their implementation, and the difficulties related to real-world operational environment.

In the following, we stay focused predominantly on large-scale research and technology development projects, which demonstrate significant technological innovations and proof-of-concept demonstrators. A common feature of these projects is that they handle large amounts of recordings collected in real-world conditions. Coping with the challenges of real-world operational conditions, these

---

3 The EC LIFE+ programme, http://ec.europa.eu/environment/life/funding/lifeplus.htm
4 The Biodiversa network, http://www.biodiversa.org/
5 The EC Structural and Investment Funds, http://ec.europa.eu/regional_policy/en/funding/
6 INTERREG Europe, http://www.interregeurope.eu/
7 The National Science Foundation (NSF) of the USA, http://www.nsf.gov/
8 The National Council for Scientific and Technological Development (CNPq), http://cnpq.br/
9 The CAPES Foundation, Ministry of Education, Brazil, http://www.capes.gov.br/
10 The EU-Brazil OpenBio project, http://www.eubrazilopenbio.eu/
11 The COOPEUS: Connecting Research Infrastructures project, http://www.coopeus.eu/

projects brought forward some key technological innovations and facilitated the implementation of ambitious technological solutions and proof-of-concept prototypes that have the potential to transform the manner in which sound-emitting species are monitored. In brief, the application scenarios covered in these projects can be summarized into three broad categories:
1. biodiversity assessment and monitoring studies (e.g. projects Automated Remote Biodiversity Monitoring Network [ARBIMON],[12] Automatic Acoustic Monitoring and Inventorying of Biodiversity [AmiBio],[13] INAU Pantanal Biomonitoring,[14] Scaled Acoustic BIODiversity platform [SABIOD][15]);
2. pest control applications in agriculture (e. g. project ENTOMATIC[16]); and
3. population control for malarial mosquitoes and other flying bloodsucking species that transmit dangerous diseases (e. g. projects Remote Mosquito Situation and Identification System [REMOSIS][17] and Anofeel[18]).

We consider these projects among others mainly because these illustrate quite well the practical significance and the social importance of technological innovation brought by computational bioacoustics. More importantly, however, these projects bring social innovation and well demonstrate the potential of automated technology when addressing a wide range of biodiversity monitoring challenges, pest control in agriculture, and risk management of diseases transmission by bloodsucking insects. The technological solutions pioneered in these projects benefit of recent advances in information technology and communications, and in some cases of the availability of hi-tech infrastructure. It is demonstrated how the overall organization of work in these projects is based on technological solutions and intelligent information processing. Finally, these projects are considered excellent examples and the present-day success stories that illustrate the applicability, potential, and limitations of contemporary methods of computational bioacoustics.

Throughout the following sections, the focus remains on computational bioacoustic methods; however, for the purpose of comprehensiveness, we also outline the overall perspective and the scope of their use in the specific

---

[12] The ARBIMON Acoustics project, http://arbimon.uprrp.edu/
[13] The AmiBio project, http://www.amibio-project.eu/
[14] The INAU Pantanal Biomonitoring project, http://www.inau.org.br/laboratorios/?LaboratorioCod=12&ProjetoCod=94
[15] The SABIOD project, http://sabiod.univ-tln.fr/
[16] The ENTOMATIC project, http://entomatic.upf.edu/
[17] The REMOSIS project, http://remosis.bg-counter.com/
[18] The Anofeel project aims at investigating the mating behaviour of malaria mosquitoes using state-of-the-art synchronized 3D video and audio recording devices.

application set-up. In this connection, we also briefly outline the objectives, functionality, methodology, and tools specific for these projects. We hope this helps in highlighting the role of computational bioacoustics and illustrating the significance of these advancements to the overall technological solution developed in each project. We also briefly mention some examples of tools and services developed for addressing specific practical needs.

8.1 The ARBIMON project[19]

The ARBIMON project is the first large-scale initiative that makes profound use of automated technology for long-term acoustic monitoring of terrestrial habitats. The long-term vision of the ARBIMON Acoustics team was to establish a research infrastructure for automated acoustic monitoring of biodiversity in support of biologists and ecologists who study spatial abundance of sound-emitting species, their population dynamics, long-term changes in various habitats, investigate biodiversity trends, and so on.

The first phase of the project, referred to as ARBIMON Acoustics, started in the year 2007. It aimed to create an automated network for data acquisition, transmission, storage, and processing. The second project phase, referred to as ARBIMON II, is focused on the establishment of a web-based platform for (semi)automated analysis of wildlife audio recordings.

8.1.1 The ARBIMON Acoustics project

The ARBIMON Acoustics project puts into operation a network of remote stations for permanent acoustic monitoring of various terrestrial habitats. This network provides a certain degree of scalability and makes possible the simultaneous acoustic monitoring of sound-emitting species at various locations, in their natural habitats, remotely, unobtrusively, and in real time (or near real time). In addition, the ARBIMON Acoustics project established a data storage repository with a relational database and certain tools for automated data processing, annotation, modelling, content mining, search of content on different levels of abstraction, and so on. In brief, the ARBIMON Acoustics biodiversity monitoring infrastructure (cf. Figure 8.1) consists of

---

[19] The ARBIMON Acoustics and ARBIMON II projects were conceived and implemented by an interdisciplinary team of researchers at the Department of Biology and the Department of Computer Science, University of Puerto Rico-Rio Piedras, San Juan, Puerto Rico, United States. The ARBIMON Acoustics project, http://arbimon.uprrp.edu/arbimon/index.php/home-acoustics

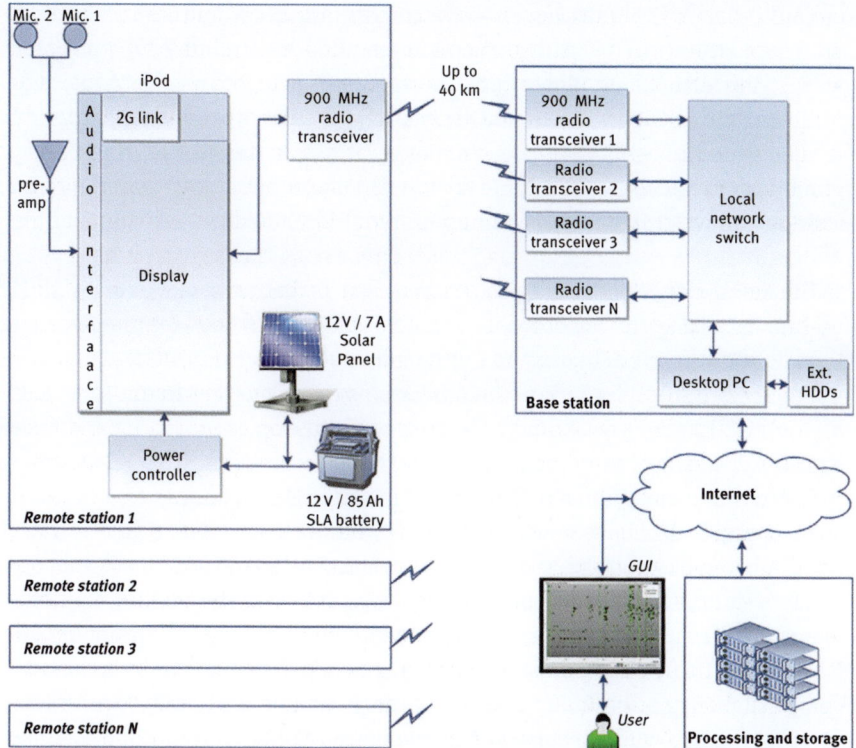

Figure 8.1: The ARBIMON Acoustics project: audio acquisition, transmission, storage, and processing.

- multiple *remote stations* deployed in various habitats that are capable of collecting and transmitting soundscape recordings;
- a *base station* for data aggregation, compression, and upload; and
- a server facility for extensive audio processing and permanent data storage.

The entire infrastructure was integrated based on commercial off-the-shelf components in order to minimize development time and reduce the overall project cost. The remote stations that form the backbone of the audio collection network are based on a commercial iPod device (Aide et al. 2013). The iPod was equipped with a stereo microphone and a custom-built application, which sets the iPod to serve as a reprogrammable audio recorder. A microphone pre-amplifier was used in order to improve sensitivity to weak sounds and thus increase the perceptive range of the *remote station*.

The commonly referred audio acquisition schedule is capturing 1 min of audio once in every 10 min, which totals to 144 one-minute soundscape recordings per

24 h (Aide et al. 2013). However, the software for audio recording that runs on the iPod device allows flexible data acquisition schedule, depending on the project's needs. The selected 1-min length of audio recordings is linked mostly to the communication channel bandwidth and energy efficiency of the remote station, and the capacity of the permanent data storage repository at the ARBIMON premises. Therefore, the ARBIMON team made the implicit assumption that sampling 1 min per every 10 min provides a fair representation of the acoustic activity near the remote station.

The main purpose of each *remote station* is to periodically record the sounds of nature and transmit the recording to the *base station*. A key point here is that the audio recordings are not stored locally but instead are transmitted wirelessly via radio communication link to the *base station*. The selected frequency channel (900 MHz), radio power, and antenna permitted transmission on a distance between 1.2 and 40 km depending on the weather conditions and density of vegetation in the direction of radio communication link.

Each *remote station* is equipped with an autonomous power source – a car battery charged by a solar panel through a voltage controller. Such a power source facilitates extended autonomy of operation, which lowers the cost for maintenance of the remote stations. All components of the *remote station*, except the microphones, solar panel, and radio antenna, are packed in a waterproof case as protection against humidity and mechanical damage. The limitations due to the specific design of the *remote station* are related to
- the range of feasible values for audio sampling frequency,
- the number of audio channels, and
- the relatively narrow bandwidth of the radio communication channel.

In stereo mode and high sampling frequency, which are considered in the particular set-up, the communication link permits only a few minutes of audio to be recorded and transmitted each hour. The audio captured by each *remote station* is transmitted to the *base station* via the 900 MHz radio communication link. At the *base station* side, multiple radio receivers are attached to a local area network (LAN) switch, with one receiver[20] for each *remote station*. A desktop PC is also connected to the same LAN switch to collects the audio transmitted by many remote stations. The primary purpose of this PC is to

---

20 In general, taking advantage of a certain time-division scheme, one radio receiver could potentially collect data of several remote stations. This is feasible only when each remote station uses the communication channel periodically (e. g. only for a few minutes per hour) and if the communication protocol allows such a functionality. Alternatively, when time-division protocol is not desirable, or when continuous audio monitoring is a valid option, each remote station requires a separate receiver at the side of the base station. Such an approach suffers from practical

- detect the arrival of a new audio recording,
- convert the audio from stereo to mono,
- compress the audio files,
- make a local copy on an external hard disk drive (HDD), and
- upload the compressed data to the ARBIMON Processing and Storage server though a wideband Internet connection.

Lossless audio compression based on the FLAC[21] encoding/decoding tools was applied on each audio file in order to reduce the requirements to the long-term data storage and for a more economical use of the Internet connection bandwidth when data are sent to the ARBIMON Processing and Storage server.

Once audio files are upload to the ARBIMON Processing and Storage server, they are archived in the permanent storage repository. The raw audio data, the audio spectrograms, or their processed versions are made available to registered users through a web-based interface. After successful login, registered users can select a specific audio recording by means of filters that allow specifying the particular subproject/location/date/time through a web-like interface, which is linked to a database. In addition, the ARBIMON Processing and Storage server provides some data management tools and tools for (semi-)automated annotation and recognition of sound events and species.

In such a way, taking advantage of the ARBIMON infrastructure and Internet, biologists and citizens have remote access to monitor acoustic environment at various locations. The ARBIMON data repository contains all recordings made since the year 2008 up to present. Besides, a web service running on the ARBIMON Processing and Storage server allows registered users to listen, annotate, and search in preselected data set, either manually or through a certain (semi-)automated tools. Some more sophisticated tools and services for the automated detection of sound events and of regions of interest (ROI) in the audio spectrogram became publicly available with the second phase of the project, referred to as ARBIMON II.

8.1.2 The ARBIMON II project

The ARBIMON II project is a follow up of the ARBIMON Acoustics project. The main objective of ARBIMON II was to establish a cloud-computing-based

---

problems when the number of radio channels increases – for instance, consider the scalability issues when one base station has to serve as a data concentrator for a few dozens of remote stations, or more.

21 The Free Lossless Audio Codec (FLAC), https://xiph.org/flac/

software platform for (semi)automated analysis of soundscape recordings and to provide convenient services to biologists and bioacousticians in support of their fieldwork. This platform, referred to as the ARBIMON II platform[22] and the Sieve Analytics web services,[23] was launched in November 2014. These tools and services make use of computational bioacoustic methods and facilitate tasks associated with the

- visualization of soundscape recordings and their audio spectrograms,
- annotation/tagging of soundscape recordings for presence/absence of certain species, and
- development of automated species-specific recognizers.

The development of an automated species-specific recognizer involves the implementation of a specific workflow, which AmiBio II platform aims to facilitate. As it is already clear from the presentation in Chapters 3, 4, and 7, this workflow requires

1. the creation of a species-specific acoustic library,
2. the development of an acoustic model and tuning the system parameters, and
3. the evaluation of the recognition performance.

Usually, the recognizer development is an iterative process, which involves several iterations of these three steps, and therefore would be facilitated by the availability of appropriate tools. Furthermore, once the recognition accuracy becomes acceptable, another tool helps to select a subset of recordings and run the recognizer on it. In brief, these tools and services rely on advanced methods of computational bioacoustics for

- semi-automated sound events annotation in support of acoustic libraries creation (cf. Chapter 4);
- audio signal parameterization based on time-frequency transformations, akin to the methods discussed in Chapter 5;
- spectrogram-based representation and processing with image processing techniques, akin to the 2-D audio parameterization outlined (Chapter 6); and
- automatic detection of ROI and sound events detection, which are based on HMM, GMM, decision trees, or other machine learning methods, akin to those discussed in Chapter 7.

In addition to the cloud-based platform, the ARBIMON II project also modernized the design of the remote stations for audio acquisition and transmission. The new

---

[22] The ARBIMON II platform, https://arbimon.sieve-analytics.com/
[23] The ARBIMON Acoustics web application, http://www.sieve-analytics.com/#!arbimon/cjg9

design is based on a contemporary smartphone with 16 GB, touch screen, and 3G and Wi-Fi connectivity. The pair of microphones was replaced with a single microphone, which slightly worsened the spatial sensitivity (in the direction backwards to the microphone). In the new design, the radio communication link is with an extended range (reportedly up to 65 km). Furthermore, the radio communication link is now optional and is considered essential only at locations where Wi-Fi and 3G wireless infrastructure is not readily available.

Due to the use of recent technology, the new *remote stations* make use of a smaller rechargeable battery (6.4 Ah) and a much smaller solar panel (14 W). The smaller dimensions of the researchable battery and other modules made it possible for all components to be placed in a compact waterproof case.[24,25] The last renders the *remote station* smaller and lighter (weights approximately ~1 kg), and thus easier to deploy and maintain. Without the solar panel, the *remote station* can be used as a portable device capable of 20 days autonomy, given that audio capturing schedule is set to recording 1 min every 10 min. The convenient user interface, the increased CPU power, and local storage in the new *remote station* provided extra flexibility and reduced the risks of information loss due to transient connectivity failure of the radio communication channel.

However, the greatest advance with respect to ARBIMON Acoustics project is that the ARBIMON II platform makes use of cloud-based services for data storage and processing, which provide convenient tools and extra flexibility of data access and analysis through a web-based user-friendly interface. Finally yet importantly, the ARBIMON II platform permits registered users to conveniently share their projects design, data, audio recording, annotations, models, and so on, and thus foster collaborations, and bring back benefit to society.

It is very important to emphasize that the ARBIMON team made a long-term investment of efforts and resources to establish and support a community of researchers that make use of computational bioacoustic methods. The community contributes to the enrichment of ARBIMON II platform with data, methods, and tools. All these are of great value and assistance to the efforts towards the implementation of scalable studies in support of global biodiversity preservation (Jahn and Riede 2013).

This vision encouraged the ARBIMON team to establish and sustain free public access to data on the ARBIMON II platform. After a free of charge subscription to the platform, everybody interested has access to all ARBIMON recordings. Furthermore, many of the external research projects hosted on the ARBIMON II

---

[24] Sieve Analytics, http://www.sieve-analytics.com/#!buy/cju7
[25] Sieve Analytics on Twitter: https://twitter.com/sieveAnalytics

platform also opted free access to their data collections and resources they created. Such an attitude for data and resources sharing (i) opens new opportunities for cross-border and cross-disciplinary collaborations, transparent, and reproducible research; and (ii) facilitates interdisciplinary biodiversity assessment studies.

Here, we need to point out that over the past decade the ARBIMON team managed to secure continuous funding, which allowed the use of contemporary methods and technology. The gradually advanced tools and services recently reached the point of commercial products in support of acoustic biodiversity monitoring[26] – remote stations, data storage, data processing tools, working environment that permits collaborative research studies, and so on. All these provided opportunities for sustainability and growth over time, which keeps ARBIMON on the forefront of advances in computational bioacoustic technology.

As a final point, we would like to clarify that we credit the ARBIMON project as the first long-term project to promote the use of automated technology for data collection, transmission, and processing, and the longest lasting among all biodiversity monitoring projects that made extensive use of automation. All this, together with the free data access, and the opportunities for collaboration it provides, makes it one of the most influential success stories of computational bioacoustics.

8.2 The AmiBio project

The AmiBio project aimed to deploy a real-time acoustic monitoring network in the area of Hymettus Mountain in Greece. The ultimate goal was to develop an automated system for biodiversity monitoring which is also capable of real-time detection of potentially dangerous events and of identifying evidence of ongoing illegal activity, such as tree felling, motocross, forest fires, and so on. Such a functionality would greatly facilitate the efforts of public organizations engaged with the protection of the natural resources of Hymettus Mountain, the firefighter brigades, and the law enforcement organizations. The AmiBio project was conceived in the beginning of the year 2008; however, the financing and therefore the actual

---

[26] In fact, the development of commercial products and services, such as recorders for all types of habitats and species, software for automated recognition of sound events, cloud-based storage and services, was pioneered by Wildlife Acoustics Inc., https://www.wildlifeacoustics.com/. Here, we credit the ARBIMON team as they developed own commercial platform, which also provides public access to own recordings and some resources shared by hosted projects.

implementation[27] of field activities at the Hymettus Mountain began in February 2010 and completed in July 2013.

The proclaimed specific goals and expected outputs of the AmiBio project were numerous and diverse, and by this reason somehow perceived as overoptimistic in the bioacoustic community, at least within the limited project duration, allocated budget, and available resources. In particular, among the expected outcomes were

- biodiversity assessment and inventorying in the Hymettus Mountain area;
- estimation of the density of animals in the monitored areas;
- monitoring and alarming about the presence or absence of rare and threatened species in inaccessible areas as well as of night-migrating birds;
- estimation of the health status of certain species based on their vocalizations;
- monitoring and alarming of specific atypical sound events, such as those related to illegal or potentially hazardous human activities, such as gun shots, tree felling, illegal races with motorcycles, and so on, and reporting to responsible authorities about an emergency situations;
- 24/7 monitoring for danger and crisis events, monitoring for natural calamity, and detection of human-induced disasters.

The achievement of these expected outcomes depended on the deployment of an automated system for acquisition, transmission, storage, detection, analysis, and visualization of acoustic and weather data, so that forest management organizations, law enforcement authorities, and firefight brigades can take prompt action when needed. This motivated the creation of an automated real-time monitoring network (Figure 8.2), which consisted of remote stations deployed in the four major types of habitats at the Hymettus Mountain, such as mid- to low- forest canopy, semi-open scrublands, and around wetlands.

---

27 The AmiBio project was implemented with co-financing of the EU LIFE+ Nature & Biodiversity programme (http://ec.europa.eu/environment/life/funding/lifeplus.htm), which provided approximately 50 % of the overall budget. The project was implemented a NATURA 2000 area: the Hymettus Mountain near Athens, Greece, which is protected as a National Park under the legislation of Greece. The interdisciplinary consortium was led by the University of Patras, Greece, and the project activities were implemented in partnership with the (i) Technological Educational Institute of Crete (TEIC), Greece; (ii) Association for Protection and Development of Hymettus (SPAY), Greece; and (iii) Zoologisches Forschungsmuseum Alexander Koenig (ZFMK), Germany. In addition, a number of project activities were supported by the (i) telecommunication operator COSMOTE, Greece, which offered its 3G wireless network at no cost for data transmission during the entire project duration; (ii) National Center for Scientific Research DEMOKRITOS, Greece, offered the ACEPT-AIR sensors and sponsored the AmiBio InfoDay in Athens; (iii) Leibniz Biodiversity Network, Germany, which sponsored the AmiBio InfoDay in Bonn. The AmiBio project website, http://www.amibio-project.eu/.

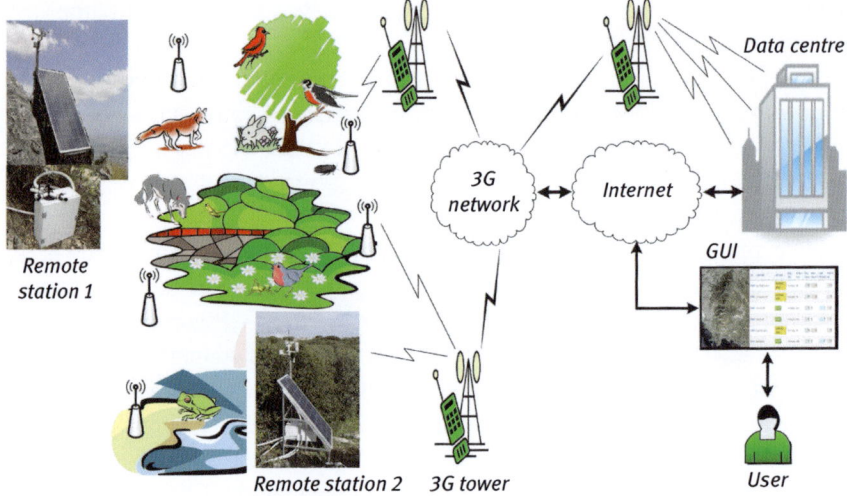

Figure 8.2: The AmiBio project concept.

The remote stations deployed in these habitats register, record, compress, and transmit audio and weather data to a remote *data centre*, where audio streams are stored permanently and analysed automatically or semi-automatically. The permanent data archival provides opportunities for long-term monitoring of biodiversity trends in the four main habitats at Hymettus. Four remote stations were deployed in each habitat type in order to provide data for statistical analysis. One additional *remote station* was deployed in the botanic garden at Hymettus. Thus, the proof-of-concept prototype network consisted of 17 remote stations. Nevertheless, the project concept and technical design permit larger number of remote stations to be deployed afterwards (on the go), and therefore AmiBio offered straightforward scalability (Ganchev et al. 2011). Such scalability provides flexibility as the network size can be adjusted to the needs of a particular plot study and deployed in subsequent increments according to the available budget and resources.

The remote stations were integrated taking advantage of off-the-shelf components, hardware modules, and software tools. Each *remote station* consisted of an *audio acquisition unit* with Wi-Fi and 3G connectivity, 4 GB of local storage capacity, 4 USB microphones, a 65 Ah car battery, a 0.6 m^2 solar panel,[28] and a voltage controller. A number of recording stations were equipped also with

---

[28] Due to the free public access to the Hymettus Mountain and its statute of a recreational area, there were restrictions imposed on the use of solar panels in certain areas – authorities requested

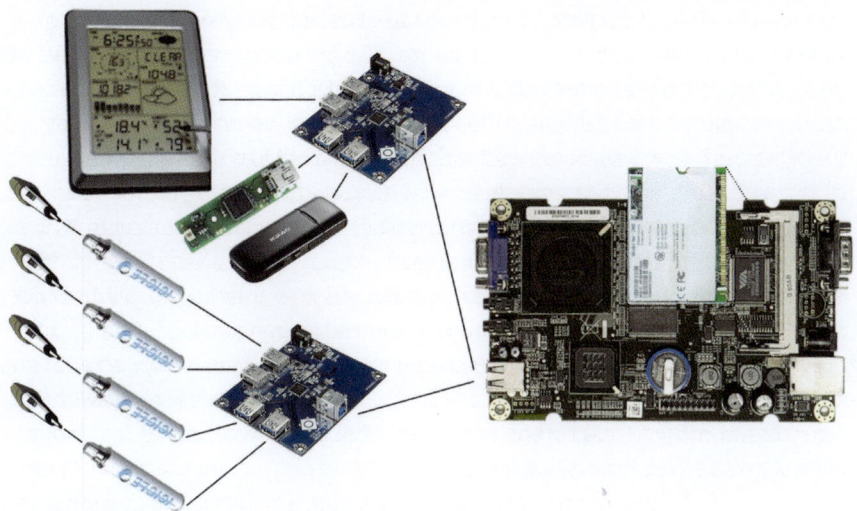

Figure 8.3: The conceptual integration diagram of the *audio acquisition unit*.

a small weather station, which registers the local microsite conditions, such as wind direction and strength, humidity, temperature, rainfall, illumination, and so on. The custom software developed by the AmiBio project, along with some open-source signal processing applications, was hosted on a small Linux-based single-board PC, which served as an *audio acquisition unit*, local processing unit, and communication unit.

The conceptual hardware integration diagram for the *audio acquisition unit* and its links to the weather station WH-1090 and 3G modem is illustrated in Figure 8.3. (The Wi-Fi radio antenna and the 4 GB data storage card are not shown.)

The main CPU and RAM resources were provided by the commercial single-board computer ALIX 3D3.[29] An 802.11a/b/g DualBand miniPCI-extension board[30] was used for implementing the Wi-Fi connectivity. The 3G connectivity was implemented via a 3G USB modem and 3G SIM card provided *as is* by the telecom operator. The 4 GB data storage card hosted the operating system, application software, and served as a temporary storage for the four audio streams and the

---

the project equipment to remain less visible to visitors. For that reason, solar panels were mounted only on approximately half of the remote stations, and for the rest the car batteries were replaced for recharge every 2 weeks. In few locations, power supply was available from nearby buildings or by the electricity network available in close proximity to the remote stations.

29 The PC Engines website, http://www.pcengines.ch/alix3d3.htm
30 The Wistron CM9-GP MiniPCI Card, http://www.mini-box.com/Wistron-CM9-GP-Atheros-miniPCI

compressed data before upload. Since the ALIX 3D3 board has only two USB ports, two 4-port USB hubs were used to accommodate the interface to a total of seven USB devices, among which are the four USB microphones, the weather station, the 3G modem, and the illumination sensor. The AmiBio project relied on low-cost electret microphones combined with low-cost ICICLE XLR-to-USB converter[31] that supports sampling frequency of 44.1 kHz and resolution of 16 bits per sample.

Three of the microphones were arranged coplanar, placed horizontally and directed at 120° one from another in order to cover the full panorama of 360°. The fourth microphone was mounted perpendicular to the three coplanar ones and pointed upwards in order to capture sound emissions from birds and bats flying over the remote stations. Some of the remote stations were equipped with an ultrasonic USB microphone,[32] capable of registering ultrasonic emissions with frequency bandwidth of 100 kHz and resolution of 16 bits per sample. The sampling frequency of 200 kHz was assumed sufficient for capturing ultrasonic emissions of all bat species inhabiting the Hymettus Mountain. The ultrasonic microphone was always positioned to point upwards, replacing the vertical microphone in the above-mentioned four-microphone set-up. The remote stations with ultrasonic microphone were deployed at locations where bat and insect acoustic activity was expected.

The four audio channels captured by the microphones of each remote station, including the ultrasonic one, were provisionally stored locally in *audio acquisition unit* – each audio stream in a separate .WAV file. The .WAV files were then compressed on the fly with the lossless FLAC audio codec.[33] The FLAC compression contributed to a reduction of the overall amount of data by more than 50 %. The compressed .FLAC files were next transmitted via the 3G wireless network to the *data centre* located at the University of Patras, that is, more than 250 km away of the Hymettus Mountain. In order to optimize the network traffic load over the 3G network between Athens and Patras, the telecom operator redirected through Internet all traffic associated with the AmiBio remote stations. The last also made possible AmiBio data streams to be delivered to any other location with little extra effort.

The *data centre* at the University of Patras hosted an FTP server, a file server, and an application server. Each *remote station* possessed an individual account on the FTP server as this facilitates the upload of recordings, organization of data, and provides flexibility of telemetry, maintenance, and opportunity for individual reconfiguration of recording set-up. For example, the independent FTP account

---

[31] The Blue ICICLE XLR-to-USB converter, http://www.bluemic.com/accessories/
[32] The DODOTRONIC Ultramic200K, http://www.dodotronic.com/
[33] The FLAC, https://xiph.org/flac/

allowed each remote station to periodically retrieve its configuration settings and upload the compressed .FLAC files without interference with the operation of the other remote stations. Next, the .FLAC files were automatically moved to the file server, which served as the permanent storage repository. The application server carried out all tasks associated with audio processing and analysis.

The GUI provided end users with functionality and tools helping to retrieve statistics, information about the operation of the monitoring network, alerts, and so on. The GUI was implemented as a web-based interface with multiple tabs. Two screenshots of a mock-up demonstrator of the AmiBio user interface are shown in Figure 8.4. These illustrate some of the key functionalities of the AmiBio real-time

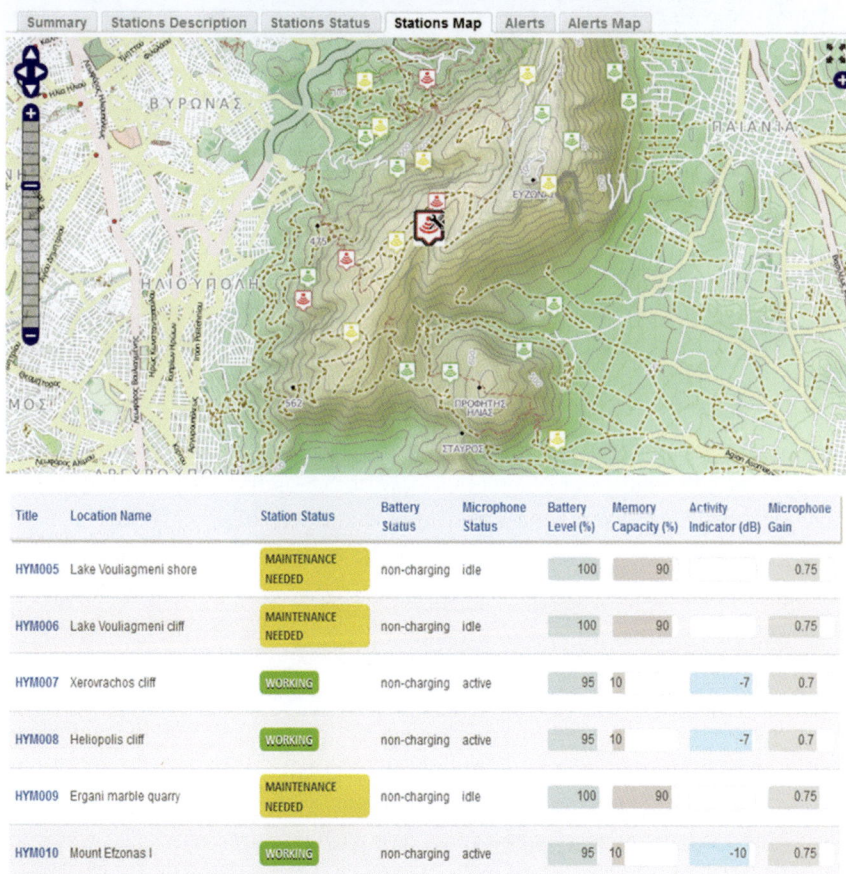

Figure 8.4: The AmiBio web-based interface. Screenshots of a mock-up demonstrator: Top: map of the AmiBio remote stations at Hymettus Mountain; bottom: remote stations status and settings.

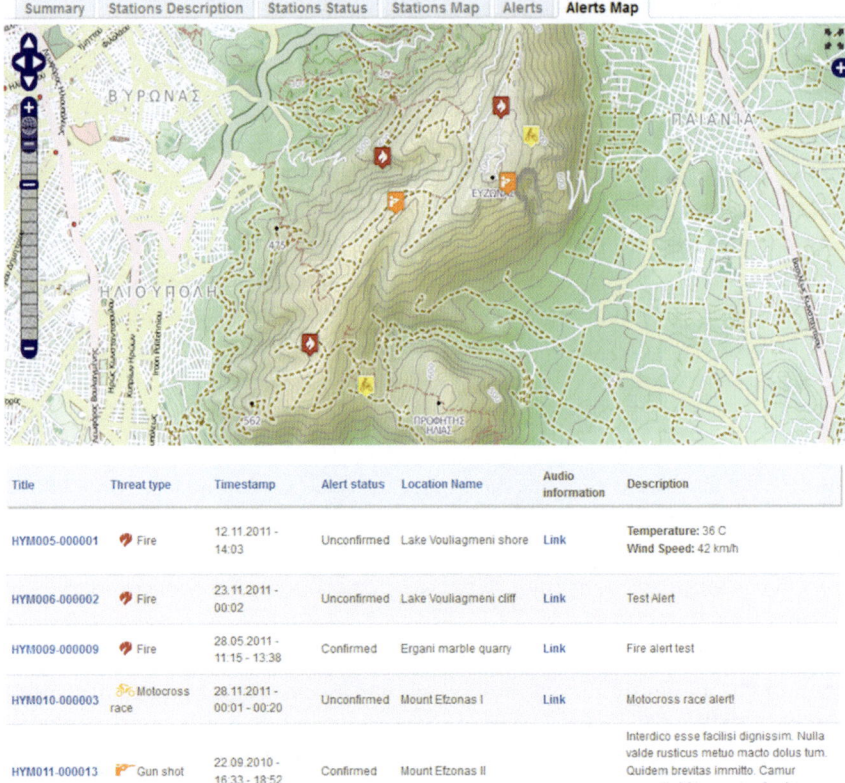

Figure 8.5: The AmiBio web-based user interface. Screenshots of a mock-up demonstrator: Top: alerts map of detected illegal activity; bottom: alerts panel.

monitoring system. Specifically, tab *stations map* (Figure 8.4, top) shows the locations of the remote stations on the map of Hymettus Mountain. The colour code of icons indicates whether the station is operational (*green*) needs attention because its resources are near critical (*yellow*) or does not operate as it requires maintenance (*red*). In such a way, just one look on the screen allows to the operator to figure out the overall operational status of the monitoring network. A more detailed information about the status and settings of remote stations is provided (after a mouse click of the station icon) through tab *stations status* (Figure 8.4, bottom). This information is updated every few minutes as the station status is obtained over the 3G network after each audio file upload. Tab *alerts map* (Figure 8.5, top) exemplifies hypothetic events of alert, which illustrate the functionality of the AmiBio system. A detailed information can be obtained after mouse click on a specific alert event – this opens the alert panel (Figure 8.5, bottom). Alerts are updated every few minutes in order to keep the information current.

The AmiBio project also developed a database and a convenient interface, which allows recordings of a certain location/date/time to be selected for visualization, annotation, processing, and so on. In total, the AmiBio project developed 32 species-specific acoustic recognizers that detect acoustic events (vocalizations) originating from a predefined set of species[34,35] (Jahn et al. 2013).

The automated detections obtained through species-specific recognizers can be used to investigate presence/absence of certain key species or to measure the overall acoustic activity at the locations where the remote stations are deployed. Such information, together with the weather and environmental condition measurements made at these locations, provides opportunities for the implementation of long-term studies that correlate the behaviour of sound-emitting species with certain climatic variables and environmental conditions.

The most significant advance in the AmiBio project was that it first demonstrated in practice the concept of scalable real-time acoustic monitoring of species in a large area, along with the real-time detection of illegal activity in the protected area of Hymettus Mountain. The main progress here is that the AmiBio project deployed a scalable system for acoustic monitoring that could be extended in a straightforward manner with a large number of remote stations, anytime. The total number of remote stations is determined only by the actual needs of a certain biodiversity study and the available budged, and not by the methodological design or some technological limitations. In such a way, one can develop a monitoring system of any size, right from the beginning, or extend it gradually over time. The deployment of additional remote stations is feasible anytime and does not require change of the overall design or the protocol of data collection, processing, or visualization. Such a scalability of real-time data collection is due to a great extent to the availability of 3G communication infrastructure and the wideband Internet connection accessible in the region of Hymettus Mountain. Such a 3G wireless infrastructure is available for over 95 % of the territory of Europe; however, it might not be readily granted at other places.

These achievements of the AmiBio project were praised by the EC LIFE+ programme[36] and AmiBio was selected among *the 13 Best LIFE Nature Projects for 2013*, and subsequently distinguished as one of *the 4 "Best of the Best" LIFE Nature Projects 2013*.[37]

---

34 AmiBio project Action E.1 report. Project Presentation. http://www.amibio-project.eu/sites/default/files/documents/pub/amibio_e1_projectpresentationreport_v1.1.pdf
35 AmiBio project Action D.8 report. After LIFE Communication Plan. http://www.amibio-project.eu/sites/default/files/documents/pub/amibio_d8_afterlifecommunicationplan_v1.0.pdf
36 The LIFE+ programme webpage, http://ec.europa.eu/environment/life/index.htm
37 Best LIFE-Nature Projects 2013, http://ec.europa.eu/environment/life/bestprojects/bestnat2013/index.htm

Regrettably, the AmiBio project was discontinued in July 2013, shortly after the LIFE+ co-financing ended – the project consortium failed to raise subsequent funding or to secure sustainability of the AmiBio monitoring system. From this perspective, the discontinuation of efforts for further advancement and elaboration of the AmiBio monitoring system (or at least securing budget for its long-term maintenance and operation) is perceived a great opportunity lost. Consequently, the consortium failed to benefit from the AmiBio system and use it to carry out valuable research on biodiversity monitoring studies.

8.3 The Pantanal Biomonitoring Project[38]

The Pantanal Biomonitoring Project started in the second half of the year 2011. Similarly to the ARBIMON and AmiBio projects, it has set the long-term objective to establish an appropriate infrastructure and resource for implementing technology-supported biodiversity monitoring in the Brazilian Pantanal. This motivated the development of
1. technological tools, which are capable of meeting the needs of automated data collection, transmission, and information processing required for such an ambitions challenge; and
2. acoustic libraries that eventually will make feasible the automated species recognition, data clustering, data content mining, and so on.

Due to the absence of readily available soundscape recordings from the Pantanal, the lack of communication infrastructure, electricity, and the difficult access to typical habitats in the area of the Pantanal, the project set a three-stage plan for achieving these objectives. The first stage, referred to as the INAU Project 3.14,[39] continued until May 2016 and was successfully implemented with the financial support provided by research funds from INAU,[40] CNPq,[41] and UFMT.[42] Certain project activities also received financing through the CsF[43] programme, the

---

[38] The Pantanal Biomonitoring Project is implemented by the Universidade Federal de Mato Grosso (UFMT), Cuiaba, Brazil, in close collaboration with the University of Bonn, Germany.
[39] The INAU Project 3.14 "Monitoring Bioindicators and Migratory Birds in the Pantanal Applied Acoustomics – a Tool for Bio-sustainability Assessment," here referred to as the INAU Project 3.14, http://www.inau.org.br/laboratorios/?LaboratorioCod=12&ProjetoCod=94
[40] Instituto Nacional de Ciência e Tecnologia em Áreas Úmidas (INAU), Brazil, http://www.inau.org.br/
[41] Conselho Nacional de Desenvolvimento Cientifico e Tecnológico (CNPq), Brazil, http://www.cnpq.br/
[42] Universidade Federal de Mato Grosso (UFMT), Cuiaba, Brazil, http://www.ufmt.br/
[43] Programa Ciência sem Fronteiras (CsF), http://www.cienciasemfronteiras.gov.br/

Brehm Fund for International Bird Conservation,[44] SESC Pantanal,[45] and other organizations in Brazil and Germany, which support biodiversity-related studies.

The main goal of INAU Project 3.14 was the study of sound-emitting animals as bioindicators for habitat quality and conservation status (Schuchmann et al. 2014). The study aimed to establish a monitoring approach in different Pantanal ecosystems in order to improve previous knowledge on local biodiversity. This is considered essential for the sustainable use of natural resources and for the implementation of appropriate conservation measures. Specifically for birds, migratory species were evaluated as potential bioindicators for aquatic and terrestrial habitats. Besides, a baseline data set was collected on the aforementioned taxonomic groups using conventional survey methods and automated recording units.

For that purpose, INAU Project 3.14 established a number of autonomous audio recording stations, which were capable of continuous recording of soundscapes. These collected sound emissions of grasshoppers, amphibians, birds, and mammals over the entire project duration. Animal communities affected by the cyclic inundation of the Pantanal were in the focus of research. In order to fulfil the above-mentioned research goals, INAU Project 3.14 established a *data centre* capable of hosting and processing hundreds of terabytes of information. The *data centre* hosted a huge collection of continuous soundscape recordings, collected in 24/7 mode at various locations of the Brazilian Pantanal. These were made with the help of a dozen of SongMeter SM2+ recorders.[46]

Between July 2012 and June 2016, the project completed the simultaneous recording of several annual cycles of continuous and schedule-driven recordings with nine or more SM2+ recorders. These recordings were collected in two distinct areas at the northern Pantanal, representative of different types of habitats. In both areas, the SM2+ recorders were equipped with two microphones. Depending on the habitat type, either both microphones were deployed at 180°, 1.7 m above the ground, or one of the microphones was replaced by a Night Flight Call (NFC) microphone[47] elevated on a pole $3\frac{1}{2}$ m above the ground. The NFC microphone is directed upwards in order to capture weak signals from birds flying high above the recording stations. Reportedly, the NFC microphone captured numerous sound

---

[44] Brehm Fonds für internationalen Vogelschutz e.V., http://www.brehm-fonds.de/
[45] Estância Ecológica Sesc Pantanal, http://www.sescpantanal.com.br/index.aspx
[46] The SongMeter product line is designed, manufactured, and marketed by Wildlife Acoustics Inc., https://www.wildlifeacoustics.com/
[47] The Night Flight Call microphone is marketed by Wildlife Acoustics Inc., http://www.wildlifeacoustics.com/en/company/news/159-wildlife-acoustics-announces-its-song-meter-sm2-night-flight-call-package-for-monitoring-migratory-birds

events associated with birds flying just over the recording station or standing on the pole with the NFC microphone. The SM2+ devices were maintained weekly and the audio recordings were transferred to the *data centre* on external HDDs. These recordings were migrated to the permanent storage repository in the *data centre* and afterwards processed manually or (semi)automatically via purposely developed software tools.

Among the most significant achievements of data collection activities was the acquisition of two full annual cycles of continuous soundscape recordings (24/7 mode) simultaneously with nine autonomous recording stations. These recording stations were deployed in a purposely designed configuration, which allows spatial coverage of a large area, sampling of various open and semi-open habitats, and detecting the direction of relocation of birds. Alike the ARBIMON and AmiBio projects, INAU Project 3.14 also implemented two additional years of schedule-based recordings with a dozen of recording stations, which recorded 15 min/h with two microphones, as well as point of interest recordings made with hand-held equipment.

The continuous audio recording, simultaneously with numerous recording stations, over period of years is a unique achievement of the INAU Project 3.14. More than 120 TB of continuous soundscape recordings were collected. This conceivably indicates a future trend in biodiversity monitoring and preservation studies – the collection and processing of big data archives of audio, images, and metadata. Such a comprehensive archive of soundscape recordings covering full annual cycles could be of great value in future comparative studies as these provide the baseline for interpretation of observations and long-standing trends.

The manual audio processing, tagging, and annotation led to the preparation of representative species-specific acoustic libraries of audio recordings and comprehensive acoustic background libraries. The acoustic background libraries represent the typical acoustic background at specific locations, where the permanent recording stations were deployed, and capture the expected acoustic conditions and their variability over daytime, weather conditions, seasons, inundation level, and so on. The acoustic library creation workflow complies with the general concept presented in Chapter 4.

Both the species-specific and the background acoustic libraries provided the resources for the subsequent development of automated acoustic recognizers, such as those discussed in Chapter 7. These are capable of detecting sounds of the target species in continuous soundscape recordings. The last is essential for the implementation of long-term studies on the life cycle and behaviour of specific indicative species, such as *Vanellus chilensis* (Ganchev et al. 2015), and for studying the influence of various environmental and human activity-related pressure on the well-being of these species.

The availability of several annual cycles of continuous recordings, collected simultaneously with nine recording stations, makes it possible to analyse acoustic activity at different locations simultaneously over a period of a year. This could help research studies to grasp the big picture of animal activity in the area, to study behavioural adaptations and other changes relevant to the inundation cycle, alterations of environmental conditions, daytime cycles, and so on.

Finally, the availability of continuous recordings made over complete annual cycles also creates opportunities to test the validity of various assumptions. For instance, researchers can investigate whether the assumption that schedule-driven recording of few minutes per hour would provide a fair estimation of acoustic activity in an area is valid. This assumption is commonly used in biodiversity studies; however, to this end it has not been validated though in-depth analysis of continuous annual cycle of soundscape recordings mainly due to the prohibitive workload required when it is carried out manually by human experts.

8.4 The SABIOD project

The SABIOD project[48] started in year 2012 and will continue until year 2017. It implements interdisciplinary research activities, which take advantage of signal processing, machine learning, ecology, biology, and physics, in order to advance computational bioacoustic technology. The main goal of SABIOD project is to develop a platform that host various tools and appropriate services which facilitate future research and management actions aiming at biodiversity preservation. Specifically, the SABIOD platform will incorporate computational bioacoustic tools and advanced information technology services for the automated detection of sound events and species, data clustering according to predefined criterion, contents indexing, tracking of individuals, handling big data, and so on.

A major distinction between SABIOD and all recent or ongoing computational bioacoustic projects is that SABIOD is not focused solely on one ecosystem type (e. g. terrestrial, river, or marine), but instead is open to cover all habitats. Such an approach provides opportunities to collect and interpret information originating from various ecosystems and simultaneously investigate in a scalable manner the numerous manifestations of the influence of human presence and activities on wildlife and biodiversity. Such an integrative approach is essential for the

---

[48] SABIOD is a CNRS Big Data Interdisciplinary project (http://www.cnrs.fr/mi/spip.php?article53), which involves partnership of over 25 entities, including laboratories from academia and research institutes, private sector companies, national parks, and associations, with the participation of over 50 researchers. The SABIOD project, http://sabiod.univ-tln.fr/

holistic assessment of human contribution to biodiversity loss, and more importantly, for gaining insights on whether (and how) management actions aiming at the protection of a certain habitat influence other ecosystems.

The SABIOD project implemented bioacoustic data collection activities in marine habitats (Mediterranean Sea), mountain habitats (Alpine), Arctic habitats (Arctic harbours), ocean (Indian Ocean), sea shores (Italy), woodland (national parks in France and Italy), tropical forests, tropical rivers (Brazil), and so on (Halkias et al. 2013; Lellouch et al. 2014; Pavan et al. 2015; Trone et al. 2014, 2015). This allows the SABIOD project to incorporate studies on a broad range of species – from whales and river dolphins to bats and birds. In such a way, SABIOD promotes a unified approach for management and processing of heterogeneous data streams captured in different types of habitats, and in different context and environmental conditions.

The SABIOD project also promoted and facilitated the public use of project data and co-organized several technology evaluation campaigns (cf. Chapter 9), workshops, and seminars, which attracted young researchers to choose a career path in the multidisciplinary research area of computational bioacoustics.

8.5 The ENTOMATIC project

The ENTOMATIC project[49] makes use of computational bioacoustics in support of integrated pest management measures against a devastating insect species, *Bactrocera oleae,* with common English name olive fruit fly, which larvae feed on olive fruits.

In attempt to control the populations of *B. oleae,* olive producers in the EU make extensive use of insecticides sprayed many times per year.[50] However, despite these efforts, olive producers regularly report over 30 % loss of production, and olive oil producers report up to 50 % financial losses due to decreased quality

---

[49] The ENTOMATIC project, entitled "Novel automatic and stand-alone integrated pest management tool for remote count and bioacoustic identification of the Olive Fly (*Bactrocera oleae*) in the field," started in September 2014 and will continue until August 2017. The project is co-financed under the FP7 SME programme BSG-SME-AG "Research for SME associations/groupings" of the European Commission. The ENTOMATIC project is led by Universidad Pompeu Fabra, Spain, and is implemented in partnership with organizations from Belgium, France, Germany, Greece, Italy, Portugal, Spain, and Turkey. Project web-site, http://entomatic.upf.edu/

[50] Recent estimates cited by the ENTOMATIC project report that the annual spending for pesticides in the eight Mediterranean countries, which produce olive oil, amounts to five billion euros.

of production and missed opportunities to deliver extra virgin olive oil.[51] The main reason for these financial loses is that *B. oleae* infestations are not detected on time, and due to lack of coordination among regional and national authorities some areas with infestation are overlooked while other are overtreated as a preventive measure. The extensive use of insecticides (i) brings health concerns, (ii) has negative effects on domestic bees and wild animal species, and thus (iii) contributes to balance shift in the ecosystem, which results in long-term harm to wildlife and humans. In the same time, the under- and overuse of insecticides brings substantial financial losses to olive producers.

The ENTOMATIC project promotes technological support to the present-day methodology for integrated pest management. The main goal is to develop an automated pest monitoring system, which will provide timely alerts about the presence of *B. oleae* adults, and spatial information about their abundance in the areas with olive trees plantations. For that purpose, the project intends to develop a purposely designed *smart trap* for *B. oleae*, data collection, and communication infrastructure, software tools for automated data processing, a geographical information system, a decision support system, and automated alert services, which will facilitate the timely treatment of infested areas. It is expected that the introduction of automated technology will contribute to a significant decrease of insecticides use and at the same time will significantly improve the efficiency of control of the *B. oleae* populations.

In brief, the ENTOMATIC pest monitoring system (cf. Figure 8.6) makes use of *smart traps* for *B. oleae*, which serve as sensor nodes integrated in a local wireless sensor network (WSN).

The data collected by each trap are relied over the WSN to a gateway, which serves as data concentrator and WSN supervisor. Each gateway is equipped with a small weather station, which logs the local weather conditions in the olive tree plantation. The gateway transmits the information aggregated from the individual *B. oleae* traps and weather data over GSM/GPRS networks to a Monitoring and Management Centre (MMC). The MMC implements the data processing and analysis, and with the help of a geographical information system (GIS) and a decision support system, makes recommendations for pesticide application, dosage, scope of spraying, and so on. These recommendations are sent as alerts to individual olive producers subscribed to the automated alert service and to the olive producers' associations.

The smart traps use baits and lures to attract adult *B. oleae* inside; however, other insect species are also attracted and enter the trap. Once an insect enters

---

[51] According to the EU legislation, extra virgin olive oil quality supposes less than 10 % of olive fruit damage.

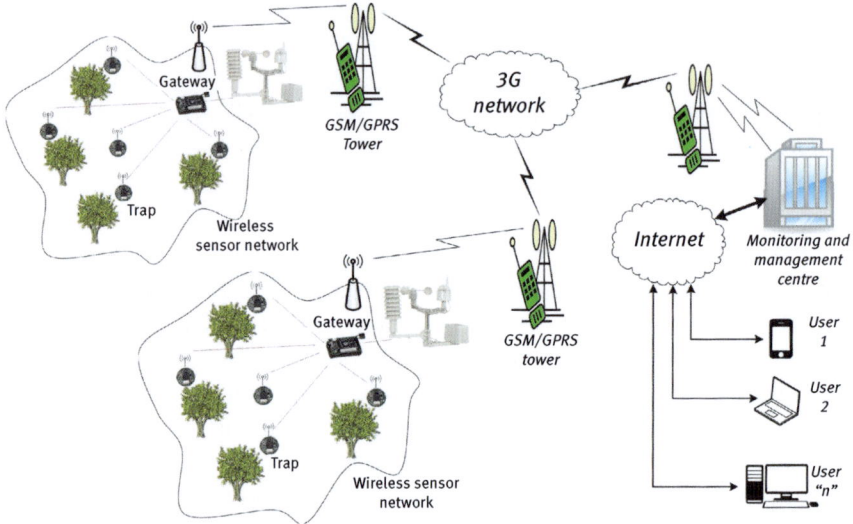

Figure 8.6: Overall concept of the ENTOMATIC project.

the trap, an optical counter registers its presence and acoustic recognizer detects whether it is *B. oleae* or another species. The automated detection process is based on the specific pattern of *B. oleae* wing sounds during flight (Potamitis et al., 2014, 2015).

The ENTOMATIC project demonstrated another success story on the application of computational bioacoustics – species-specific automated acoustic detection of insects was embedded in a real-world application for integrated management of pest populations.

8.6 The REMOSIS project[52]

The project REMOSIS aims to advance, validate, and commercialize a new smart trap for mosquitoes. It builds on an existing line of mosquito traps[53] with remote

---

[52] The REMOSIS project started in February 2016 and will continue until January 2018. REMOSIS is co-financed by the EC under the H2020 programme – topic FTIPilot-1-2015 "Fast Track to Innovation Pilot." The project is coordinated by BIOGENTS AG (Germany) and implemented in collaboration with IRIDEON SL (Spain) and Technological Educational Institute of Crete (Greece). The REMOSIS project web-site, http://www.remosis.eu/

[53] The BG-Counter (http://www.bg-counter.com/) and BG-Sentinel (http://www.bg-sentinel.com/) product lines are designed, manufactured, and marketed by BIOGENTS AG, http://www.biogents.com/

monitoring capability, which will receive new functionality for optoacoustic identification of mosquito species (Potamitis and Rigakis 2016a, 2016b). The new trap will count and automatically identify several dangerous tiger mosquito species, such as the *Aedes albopictus* and *Aedes aegypti*. These and other mosquito species are known to transmit yellow fever, dengue, Zika virus, and various mosquito-borne pathogens, which pose threat to human health.

Mosquitoes are recognized as the most dangerous creatures, with respect to disease spread and resultant human mortality. In the past decades, mosquito population control was of great concern mainly due to malaria,[54] yellow fever, and dengue.

To this end, various methods for control of the mosquito's populations were developed, among which are
1. reducing mosquito habitats,
2. spraying with insecticides,
3. sterile-insect methods, and so on.

However, to a great extent the success of these methods depends on accurate and up-to-date information about the spatial distribution of mosquito populations. In this regard, the REMOSIS project aims to facilitate the quick availability of data collected by the smart traps at various locations. Data will be uploaded wirelessly to a cloud server. IoT technology will contribute to a wider availability of data collected in different contexts for a more precise risk assessments analysis.

The project is expected to contribute towards improved control of mosquito-borne diseases, reducing the cost due to eliminating the need of manual human inspection, and focused insecticide use for reducing the negative effects on environment. The smart trap and cloud services will support future planning and surveillance actions with improved accuracy and promptness. Earlier detection and localization that is more precise will provide more opportunities for mosquito population control, which is expected to facilitate the reduction of risks associated with outbreaks of infectious diseases.

Concluding remarks

For decades, it was apparent that computational bioacoustic methods are in demand when it comes to scalability of biodiversity studies – simply the manual processing of vast amount of audio recordings is not feasible. The ARBIMON

---

[54] For instance, the ongoing project AvecNet, "African Vector Control: New Tools" – a 15.4 MEuro project aimed at malaria control, co-financed under the EU FP7-HEALTH programme. The AvecNet website, http://avecnet.eu/

project was the first long-term effort that persistently invested in this direction and made a bold step towards scalability of biodiversity studies. The idea for automated data collection, transmission, and processing implemented in the ARBIMON project was reused in various subsequent projects.

The AmiBio project, taking advantage of existing communication infrastructure, such as 3G wireless networks and wideband Internet connection, for a first time demonstrated a fully scalable network of remote recording stations, which also provided real-time acoustic monitoring for the automated detection of certain illegal and potentially dangerous activities. The availability of high-throughput communication infrastructure made the AmiBio biodiversity monitoring network straightforwardly extendable to the needs of biodiversity studies of different scale, theoretically even to hundreds or thousands of remote stations.

The Pantanal Biomonitoring Project also created infrastructure and resources that contribute towards long-term biodiversity monitoring. Its initial stage, referred to as INAU Project 3.14, was the first project to implement continuous recording of soundscapes in 24/7 mode, simultaneously with nine recordings stations, for over 2 years and thus providing many annual cycles of recordings in the area of northern Pantanal. The design plot and the deployment plan of the remote stations enabled spatial coverage and sensitivity to the gradient of sound source relocation. In this initial stage, the Pantanal Biomonitoring Project collected over 120 TB of continuous soundscape recordings, which permits the implementation of long-term studies simultaneously at different locations in the area of northern Pantanal.

The SABOID project is a nice example of an ambitious coordinated effort to create resources and technological tools in support of scalable biodiversity studies in diverse ecosystems (both terrestrial and underwater). It is the first project to consider such diversity of habitats and species, including marine mammals, river dolphins, birds, bats, and so on. This was achieved by coordinating and supporting various subprojects and activities, implemented by the large SABOID consortium. Furthermore, the SABIOD project encouraged various data collection campaigns and coordinated technology evaluation efforts. In this regard, the SABOID project co-organized and supported with recordings various public technology evaluation campaigns, among which are the LifeCLEF Bird[55] 2015 task and the NIPS4B[56] challenge (cf. Chapter 9).

The ENTIMATIC project is among the first projects that made use of computational bioacoustic technology in support of pest population's control. The

---

[55] The LifeCLEF-2015 Bird task, http://www.imageclef.org/lifeclef/2015/bird
[56] The Neural Information Processing Scaled for Bioacoustics: NIPS4B-2013, http://sabiod.univ-tln.fr/nips4b/

project had strongly applied focus and developed a prototype technology, helping to mitigate the loss of olive production in Mediterranean countries.

The REMOSIS project aims at optoacoustic detection of disease-transmitting mosquitoes. It is among the first large-scale efforts to use computational bioacoustic methods in order to provide a scalable solution that will eventually permit tracking and control of disease-transmitting mosquito's populations.

These projects target different goals and thus make use of different approaches and technology in order to deliver the desired functionality. However, all these projects have something in common, namely they bring innovation and incremental advances to computational bioacoustic methods. Furthermore, these projects proliferate the use of advanced technology for solving a wide range of real-world problems, such as biodiversity assessment studies, presence/absence studies of key animal species, pest population control applications, evaluation the spread of disease-transmitting mosquitoes, and so on.

Although the projects outlined in the present chapter represent few facets in the vast scope of computational bioacoustics, and thus do not encompass all kinds of applications, these are illustrative about the present state-of-the-art and the technology trends in the years to come. These projects demonstrate quite well what is achievable at the present stage of technology development and the current technological limitations. Furthermore, these recent projects indicate the shift of technology development trends towards

1. handling large amounts of data collected through various dissimilar types of devices in uncontrolled recording set-up;
2. use of crowd-contributed data which are captured with amateur-grade equipment; and
3. automated processing of multimodal heterogeneous data streams.

The advance of information technology and the wider availability of communication infrastructure and Internet-based cloud storage and services will definitely facilitate the inauguration of more ambitious and multifarious projects, which make use of real-time continuous monitoring technology in a scalable manner, in order to provide a broader coverage of monitoring in different ecosystems.

With the advance of computational bioacoustics and artificial intelligence methods, we anticipate the emergence of scalable fully automated systems that are able to analyse, disambiguate, and interpret big amounts of heterogeneous data. These will serve well as fundamental building blocks in expert systems, which embed multidisciplinary knowledge about climate, biodiversity, and good practices in the management of natural habitats. Such a technological support would facilitate improved planning activities and would contribute towards a more efficient management of protected areas and biodiversity preservation. It is important that management actions stay tuned to the local specifics of the areas, and in the same time account for the global consequences over extended periods, and the undesired effects these might cause to other habitats and ecosystems. Such advanced expert systems are anticipated to become efficient advisors, which provide recommendations and an effective decision support in the planning of measures aimed at global biodiversity preservation. This is conditioned on the availability of efficient and effective computational bioacoustic methods, which offer opportunities for scalable biodiversity monitoring.

9 Useful resources

Introduction

Summarizing the exposition in previous chapters, here we complement the discussion on computational bioacoustic tasks (cf. Chapter 3), methods, and applications with links and notes to some useful resources that were contributed by the bioacoustic research community.[1] Specifically, among these are
1. numerous audio collections of animal sounds (Section 9.1);
2. various publicly available software tools and services (Section 9.2);
3. online community support through mailing lists, online Q&A archives, and FAQs (Section 9.3); and
4. scientific conferences and workshops, training courses, and information campaigns oriented towards the general public (Section 9.4).

In addition, we highlight the importance of international collaboration and worldwide technology evaluation campaigns, which facilitate and speed up the advances of computational bioacoustic methods. Specifically, besides the scientific conferences, symposia, training courses, seminars, joint research missions, and technology development projects, working meetings, and so on, a number of worldwide technology evaluation campaigns were launched in support of technology development activities. These technology evaluation campaigns (i) are open to everybody who are interested; (ii) aim to foster joint activities; and (iii) establish common ground for comparison of concepts, methods, and technological tools. Among these, in Section 9.5 we only mention some recent competitions (Frommolt et al. 2008; Spampinato et al. 2012; Glotin et al. 2013c; Goeau et al. 2014; Joly et al. 2015; CLEF- 2016) as they provided an excellent support to the promotion of scalable computational bioacoustic methods. In fact, these campaigns promoted a purposely designed experimental set-up, which guided research efforts towards specific application scenarios and which fostered the development of interdisciplinary methods for an automated acoustic recognition of a large number of species and operation in real-world environment. Some of these campaigns (e. g. LifeCLEF) became annual, and thus provide strategic backing to the evaluation of scalable methods and the development of robust technological solutions. It is important to emphasize that these technology

---

[1] Further information about new resources, recent and forthcoming events is available on the home page of this book (http://tu-varna.bg/aspl/CBbook/) and on the community-supported webpages devoted to bioacoustics and computational bioacoustics.

DOI 10.1515/9781614516316-009

evaluation campaigns already attract a wide international participation, including leading research groups, and institutions engaged with interdisciplinary research in the area of bioacoustics and biodiversity studies.

9.1 Online audio data sets and resources

During the decades of data collection and field research studies, the bioacoustic community accumulated millions of sound recordings of insect, amphibian, avian, and mammalian species. Besides the extensive sound collections belonging to academia, research institutes, and natural history museums, hundreds of thousands of audio recordings were captured by people who are not associated with specific research projects. Among the most active contributors are birdwatchers and wildlife devotees, who regularly spend time in nature and have hobbies related to collection of wildlife data. In the following, we briefly mention some animal sound collections, which are well known in the bioacoustic community. Here, we do not intent to enumerate all significant animal sound archives or provide a comprehensive list, but we want to point out that such resources are available all over the globe, and more importantly, many of these are reachable online through dedicated websites. These structured and tagged sound archives constitute a key resource on which the development of computational bioacoustic methods and technology depends.

9.1.1 The xeno-canto repository

The *xeno-canto* repository[2] was established in the year 2005 as a citizen science project, which is open to anybody interested to contribute. Currently, *xeno-canto* receives bird sound recordings from both professional ornithologists and other individuals interested in nature, and is the richest community-contributed resources of bird sounds. To this end, over 3,000 contributors uploaded recordings to *xeno-canto*. The repository presently contains over 310,000 recordings of over 9,500 bird species. Besides the audio recordings and taxonomic information, the *xeno-canto* also offers audio spectrograms of these recordings, discussion forums, frequently asked questions, repository of research articles, and so on.

In the recent years, the *xeno-canto* collection of bird sounds attracted the attention of the computational bioacoustic community. This is mostly due to

---

[2] The Xeno-canto Foundation (Stichting Xeno-canto voor natuurgeluiden) receives financial support from Naturalis Biodiversity Center of the Netherlands. The *xeno-canto* repository website, http://www.xeno-canto.org/

the public technology evaluation campaigns BirdCLEF-2014,³ BirdCLEF-2015,⁴ and BirdCLEF-2016,⁵ which were entirely based on bird recordings from the *xeno-canto* repository. Specifically, in the BirdCLEF-2014 evaluation, the experimental set-up included multiple recordings of 500 bird species from Brazil, and in BirdCLEF-2015 and BirdCLEF-2016 the number of bird species was increased to 999.

9.1.2 The Borror Laboratory of Bioacoustics Sound Archive

The Borror Laboratory of Bioacoustics Sound Archive[6] was established in the year 1948 and is among the oldest active bird sounds' collections that still gather data. It currently contains over 40,000 recordings of animal sounds. More than 27,500 of these are recordings of over 3,000 bird species, which were recorded in North America and the Neotropics but the collection also included recordings from Africa, Australia, and Southeast Asia. Unfortunately, only a subset of these recordings is digitized and available online.

9.1.3 The Macaulay Library

The Macaulay Library[7] archive is among the largest sound collections of animal sounds. It contains ca. 175,000 sound recordings, which are estimated to cover about 75 % of all bird species. The sound archive also contains recordings of insect, fish, frog, and mammal (terrestrial and marine) sounds. Besides sounds, the Macaulay Library also contains photos, videos, and bird spot records. In total, there are over 50,000 videos and over 330 million bird spot records. The earliest video clip is dated back in the year 1929, which is the first record in the archive. Unfortunately, sound recordings are not freely downloadable but can be requested under a license, which depends on the purpose of their use.

9.1.4 The British Library Sound Archive

The British Library Sound Archive was established in the year 1969 and currently contains sound recordings of over 10,000 species of birds, mammals, amphibians,

---

[3] The LifeCLEF 2014 Bird task, http://www.imageclef.org/2014/lifeclef/bird/
[4] The LifeCLEF 2015 Bird task, http://www.imageclef.org/lifeclef/2015/bird/
[5] The LifeCLEF 2016 Bird task, http://www.imageclef.org/lifeclef/2016/bird/
[6] The Borror Laboratory of Bioacoustics Sound Archive, http://www.flmnh.ufl.edu/bird-sounds/
[7] The Macaulay Library sound archive is a publicly supported project of the Cornell Lab of Ornithology, (http://www.birds.cornell.edu/). The Macaulay Library, http://macaulaylibrary.org

reptiles, fish, and insects, collected at various locations around the world. However, coverage is not uniform and sounds from the UK and Europe are more abundant. The sound collection is comprised primarily of wildlife animal sound recordings collected in their natural habitats. However, there is a tiny portion of recordings with sounds of domestic animals and animals in captivity. Since the year 2004, a portion of the sound archive is available online through the British Library Sound Archive portal[8] for educational and commercial purposes. A subset of approximately 4,000 recordings of animal sounds and soundscapes is available for listening to everybody interested.

9.1.5 Sound Library of the Wildlife Sound Recording Society

The Wildlife Sound Recording Society[9] was founded in 1968 and is among the oldest and largest wildlife recording societies in the world, with around 300 members from countries of most continents. Each year the society publishes a number of CDs with animal sounds and soundscape recordings, which are commercially available[10] to non-members. MP3 compressed sound files of some recordings are also available through the society website. Among other benefits, society members receive four CDs included within their membership subscription.

9.1.6 The BioAcoustica repository

The BioAcoustica[11,12] is an open-access online audio repository, which contains wildlife recordings of insect, amphibian, avian, and mammalian species. Besides sounds, metadata, and taxonomic information, the BioAcoustica repository also provides tools for crowdsourced annotation of recordings, where contributors can edit recordings to remove noisy portions of the signal and if they like to introduce voice comments (Baker et al. 2015). At present, the BioAcoustica sound analysis tools allow researchers to visualize waveforms, audio spectrograms, and carry out dominant frequency analysis.

---

8 The British Library Sound Archive, http://www.bl.uk/collection-guides/wildlife-and-environmental-sounds/
9 The Wildlife Sound Recording Society, http://www.wildlife-sound.org/
10 The Wildlife Sound Recording Society store, http://www.wildlife-sound.org/wsrs-store.html
11 BioAcoustica: Wildlife Sounds Database, http://bio.acousti.ca/
12 Baker et al. (2015). BioAcoustica: A free and open repository and analysis platform for bioacoustics. Database: bav054. doi:10.1093/database/bav054

9.1.7 The Animal Sound Archive at the Museum für Naturkunde in Berlin

The Animal Sound Archive[13] (Tierstimmenarchiv) at the Museum für Naturkunde in Berlin contains ca. 120,000 sound recordings of over 2,500 species of invertebrates, fishes, reptiles, amphibians, birds, and mammals. Currently, over 16,000 of these recordings are available online through the archive website to anybody interested. The Animal Sound Archive also provides reference animal vocalizations for over 270 European species in support of the development of tools for automated recognition of their sounds. The aims and purpose of this archive are discussed in Frommolt et al. (2006).

9.1.8 Australian National Wildlife Collection Sound Archive

The Australian National Wildlife Collection Sound Archive[14] contains over 60,000 recordings of invertebrates, amphibians, birds, and mammals mostly from Australia but it also contains sounds from European, North American, South American, and New Guinean wildlife. Some of these sound recordings date back to the 1950s, which provides useful information for comparative studies. Some portions of the sound recordings are made available through the Atlas of Living Australia website.[15]

9.1.9 The Western Soundscape Archive

The Western Soundscape Archive[16] offers sound recordings contributed by volunteers, governmental agencies, and conservation groups. The sound collection covers over 700 amphibian, reptile, bird, and mammal species, recorded in their natural environments. The geographic coverage of this collection encompasses the western states of the USA, including Arizona, California, Colorado, Idaho, Montana, Nevada, New Mexico, Oregon, Utah, Washington, Wyoming, and Alaska.

---

[13] The Animal Sound Archive (Tierstimmenarchiv) at the Museum für Naturkunde in Berlin, http://www.animalsoundarchive.org/, http://www.tierstimmen.org/en and http://www.tierstimmenarchiv.de/
[14] The Australian National Wildlife Collection Sound Archive, http://www.csiro.au/en/Research/Collections/ANWC/About-ANWC/Our-wildlife-sound-archive
[15] The Atlas of Living Australia, http://www.ala.org.au/
[16] The Western Soundscape Archive, http://www.westernsoundscape.org/

9.1.10 The DORSA Archive

The Digitized Orthoptera Specimens Access (DORSA)[17] contains ca. 4,000 sound records of Orthoptera songs, which are linked to voucher specimens. In total, the collection contains 2,229-type specimens, which are kept in German museums. Besides sounds, the collections include geographic information, 25,000 images, illustrations of type specimens, and so on.

9.1.11 The Nature Sounds Society

The Nature Sounds Society[18] aims to encourage the preservation, appreciation, and creative use of natural sounds. Its activities are oriented towards promoting education in the technological, scientific, and aesthetic aspects of sounds on nature. The society webpages provide links to other websites with recordings published by its members.[19]

9.1.12 The Acoustics Ecology Institute

The Acoustics Ecology Institute website provides a short description of various sound archives beyond what was mentioned here. The interested reader may want to follow the short outline and links to sound libraries[20] and long list of links to other sound-oriented sites with soundscape recordings.[21]

9.1.13 Sound resources not listed here

In addition, the IBAC website[22] provides a long list of links to sound collections and research teams working in the area of bioacoustics. There are numerous

---

17 The DORSA (Digitized Orthoptera Specimens Access), A "Virtual Museum" of German Orthoptera Collections, http://bioacoustics.eu/index.html
18 The Nature Sounds Society, http://www.naturesounds.org/
19 The Nature Sounds Society, recordings published by society members, http://www.naturesounds.org/recordings.html
20 The Acoustic Ecology Institute, list of sound libraries, http://www.acousticecology.org/recordings.html
21 The Acoustic Ecology Institute, links to sound oriented websites, http://www.acousticecology.org/soundscapelinks.html
22 International Bioacoustics Council (IBAC), Bioacoustics links, http://www.ibac.info/links.html

other collections, which are well known in the bioacoustic society,[23] among which are the Florida Museum Bioacoustic Archives,[24] The Fitzpatrick Bird Communication Library,[25] Fonozoo,[26] Biblioteca de Sonidos Aves de México,[27] Veprintsev Phonotheca of Animal Voices, and so on, which we do not outline here due to space limitations. Besides, there are numerous organizations which provide commercial CDs with sounds of nature and processed recordings of wildlife, and websites[28,29,30] that provide animal sounds for download without further details about recording and taxonomic information (in the form of sound effects, sounds of nature, etc.). In addition, few case studies outlining the establishment and functioning of sound libraries in Mexico, Colombia, and Brazil are offered in Ranft (2004).

9.2 Software tools and services

Over the past decades, the bioacoustic community made use of various software tools[31,32] for audio visualization, annotation, manipulation, processing, analysis, recognition, and so on. These and other similar tools for acoustic recognition of species typically rely on audio libraries, akin to those discussed in Chapter 4, and of audio analysis and event recognition methods akin to those discussed in Chapters 5–7. Over the years, the bioacoustic community extensively used and in certain cases developed further and enhanced few of these tools,[33,34,35,36] while

---

[23] Special (Poster) Expositions of animal sound archives at the International Bioacoustics Congress
[24] The Florida Museum Bioacoustic Archives, https://www.flmnh.ufl.edu/index.php/birds/home/
[25] The Fitzpatrick Bird Communication Library, Transvaal Museum, Pretoria,
[26] The Fonozoo website, links, http://www.fonozoo.com/eng/enlaces.php
[27] Biblioteca de Sonidos Aves de México, http://www1.inecol.edu.mx/sonidos/menu.htm
[28] Freesound, http://freesound.org/
[29] Free Sound Effects, https://www.freesoundeffects.com/free-sounds/animals-10013/
[30] Listening Earth, www.listeningearth.com/LE/
[31] List of Bioacoustics Software, https://en.wikipedia.org/wiki/List_of_Bioacoustics_Software
[32] List of Sound Analysis Software for Bioacoustics compiled by Steven Hopp, http://science.ehc.edu/~shopp/sound.html
[33] Raven, Interactive Sound Analysis Software, http://www.birds.cornell.edu/brp/raven/Updates/index.html
[34] WildLife Acoustics Kaleidoscope, https://www.wildlifeacoustics.com/products/kaleidoscope-software
[35] WildLife Acoustics SongScope, https://www.wildlifeacoustics.com/products/song-scope-overview
[36] Avisoft SASLab Pro, http://www.avisoft.com/soundanalysis.htm

other useful tools remained less popular. The last one is not always in direct relation to the functionality offered by these tools and is influenced by numerous other usage aspects and cultural trends. Among these are
1. *Institutional policy*: institutional culture, what tools are recommended, for example, is the tool in-house built or not,
2. *Accessibility*: is the tool freeware or it can be purchased under a licence, price;
3. *Operating system*: is the required operating system free for research use or licensed;
4. *Marketing*: is the tool actively promoted for public use;
5. *Documentation*: quality of the user manual, training opportunities;
6. *Acceptability*: convenience and intuitiveness of the user interface;
7. *Support*: the availability of an online community support, error reporting, updated versions, and so on.

These and other aspects are not always considered when new tools and services are developed. This is because many tools are made available for public use by research teams or individual researchers who created these as a part of a research project or study, and not with the intention to develop and market a successful commercial tool. By following practical considerations, researchers tend to approach the software development giving priority to convenience and quickness, and thus, such tools are provided *as is*, with incomplete documentation, limited support, run on a single operating system, and so on.

At present, most of these tools and services offer a narrow range of functionality, which is sufficient to illustrate the potential of technology. However, this is not always sufficiently attractive to convince bioacousticians to adapt traditional methods and data processing workflow (predominantly based on manual labour) and make the workflow utterly technology oriented. This is because automated tools offer certain basic functionality; however, these are prone to errors and various biases. Furthermore, their performance is not always predictable especially when operate on soundscape recordings – recognition accuracy fluctuates significantly depending on the noise floor, presence of interferences, and changes in the weather conditions and environment.

Still, many of these automated tools offer valuable assistance, help in reducing the time and effort needed for search of a predefined sound event type or species, and provide excellent support to some annotation or tagging of recordings tasks. The greatest challenges and limitations remain linked to (i) providing a reliable operation under adverse noise conditions and (ii) the lack of comprehensive coverage of species, a prerequisite in biodiversity assessment studies. The former refers to the relatively low robustness when working with noisy soundscape recordings, and the latter to the quite limited coverage of ready-to-use

species-specific models. Another issue is that currently technological tools require significant investment of time for training and preparation of data. One way to overcome many of these problems is to create a software platform
1. with convenient user interface that combines many tools;
2. which will be used by a large number of researchers and practitioners, who are inclined to share their work: recordings, data sets, models, results;
3. which provides collaboration fostering tools and opportunities users to contribute and share on their will own content and tools; and
4. which provides support and discussion forum aimed at mutual assistance between users, and so on.

Fortunately, in the recent years the integration of software tools has gained momentum. Tools implementing species and sound events recognition and other functionality, such as data indexing, search of recordings based on metadata or annotations, and so on, were embedded into integrated software platforms[37] that aim to provide tools and services in support of bioacoustic research activities. Among these activities are tasks related to the creation of presence/absence maps, diagrams of daily/monthly/seasonal activity patterns, and so on. Some of these software platforms already reached maturity that allows commercialization of services, most often bundled in a package with audio recording equipment, cloud storage and computing resources, geographic information systems, maps, and other products and services.

Most often, the acoustic event recognition tools needed are already integrated in the software platform but the users have to
1. provide the annotated training libraries or to annotate data directly in the environment of the specific platform,
2. build appropriate acoustic models, and
3. create own recognizer for each specific species or for each sound event of interest.

These software platforms already attract the attention and raise the expectations of researchers. This is mainly due to the potential behind these tools and services to automate and speed up the data acquisition, the search of recordings based on time location and project, the management of recordings, and so on. However, at present, these tools and services have not become widely used in biodiversity assessment studies. The last is mostly because of the prohibitive workload required for the preparation of prerequisite annotated audio libraries for multiple species and the respective species-specific recognizers.

---

[37] The ARBIMON II platform, https://arbimon.sieve-analytics.com/

We hope that this will soon change, and the current generation of PhD students and young researchers will embrace the challenge to actively use these software platforms. Sharing species-specific acoustic libraries and models, contributing to further development of the functionality and resources needed for the implementation of scalable biodiversity studies would bring benefits to individual researchers and their institutions, and more importantly, a significant social impact and a wider public support.

9.3 Online information

Established a century ago, bioacoustic research studies already enjoy various benefits due to
1. the accumulation of knowledge, methods, and audio resources (recordings);
2. the advances of communication and information technology methods, tools, and services; and most notably
3. the information communication opportunities offered by Internet.

Among these, the establishment of online tools and services that facilitate the communication and the exchange of information, the sharing of experience, good practices, audio resources, contribution of software tools and models, and so on, are of key importance. These help fostering international and interdisciplinary collaboration, and also providing more support and opportunities to young people who are interested in joining the bioacoustic community. Some useful tools are as follows:
1. *Webpages.* The bioacoustic community supports numerous webpages, among which crowd-contributed information pages,[38,39] home pages of academic research laboratories, nongovernmental institutions, and commercial entities with activities associated with bioacoustics and biodiversity studies.
2. *Emailing lists.* The bioacoustic emailing list[40] <bioacoustics-l.cornell.edu>, hosted at the Cornel University, is widely known and commonly used for posting announcement of new research studies, research positions, job offers, and discussions on specific topics. Another important email list[41] is <list@ibac.info>, set by the International Bioacoustics Council (IBAC), which is focused on topics related to the IBAC events.

---

38 Bioacoustics page at Wikipedia, https://en.wikipedia.org/wiki/Bioacoustics
39 List of Bioacoustics Software, https://en.wikipedia.org/wiki/List_of_Bioacoustics_Software
40 The email list <bioacoustics-l@cornell.edu> is archived at The Mail Archive repository, https://www.mail-archive.com/bioacoustics-l@cornell.edu/maillist.xml
41 The IBAC mailing list subscription, http://www.ibac.info/contacts.html

Other important resources are online FAQs, bibliographies, presentations, and short communications presented at scientific events, workshops, and conferences, articles published in scientific journals, among which *Bioacoustics*,[42] *Soundscape*,[43] *JASA*,[44] *PLoS ONE*,[45] *Applied Acoustics*,[46] *IEEE/ACM Trans. on ASLP*,[47] *Ecological Informatics*,[48] and so on.

9.4 Scientific forums

The bioacoustic research community organizes numerous scientific events, such as conferences, symposia, workshops, and other important scientific meetings, where the newest developments are reported. The major events among these are the IBAC Meetings,[49] organized by the IBAC, and held every second year since the early 1970s.

Each year, there are numerous other conferences, workshops, meetings, project-supported and other events[50] devoted to bioacoustics, or including sessions on bioacoustics. Among these are the various conferences of the World Forum for Acoustic Ecology, such as the Ecoacoustics Congress, Acoustic Ecology Conference, The Symposium on Acoustic Ecology, and International Conference of the World Forum for Acoustic Ecology, which also receive submissions that overlap with the topics of bioacoustics.

Furthermore, scientific forums with broader research areas, such as those devoted to signal, speech, and audio processing, regularly organize special sessions, workshops, and receive submissions reporting developments in the area of computational bioacoustics. Among these, the largest and most influential are

---

[42] *Bioacoustics: The International Journal of Animal Sound and its Recording*, http://www.bioacoustics.info/
[43] *Soundscape: The Journal of Acoustic Ecology*, http://wfae.net/journal/index.html
[44] The *Journal of the Acoustical Society of America (JASA)*, http://scitation.aip.org/content/asa/journal/jasa
[45] The *PLoS ONE* journal, http://journals.plos.org/plosone/
[46] *Applied Acoustics*, http://www.journals.elsevier.com/applied-acoustics
[47] *IEEE/ACM Transactions on Audio, Speech, and Language Processing*, https://signalprocessingsociety.org/publications-resources/ieeeacm-transactions-audio-speech-and-language-processing/ieeeacm
[48] *Ecological Informatics: An International Journal on Ecoinformatics and Computational Ecology*, http://www.journals.elsevier.com/ecological-informatics
[49] History of the IBAC Meetings, http://www.ibac.info/meetings.html
[50] List of scientific events related to bioacoustics, http://www.ibac.info/other_meetings.html

1. the IEEE International Conference on Acoustics, Speech and Signal Processing (ICASSP);[51]
2. INTERSPEECH organized by ESCA;
3. EUSIPCO organized by EURASIP; as well as
4. conferences of the machine learning community and the IEEE Signal Processing Society (uLearnBio,[52] ICML,[53] MLSP,[54] NIPS[55]), and so on.

9.5 Technology evaluation campaigns

As accentuated earlier, we highly credit the worldwide technology evaluation campaigns organized by the bioacoustic community. Among these, we briefly mention few recent technology evaluations that focus on the automated acoustic recognition of a large number of species, noisy field recordings, or recordings made in uncontrolled conditions. The scale of difficulty set in such an experimental set-up requires handling large-scale acoustic recognition of species and entails the establishment of scalable methods and development of robust technological solutions that operate well on real-world recordings. Such challenges align the research and technology development efforts of all participants in the competition and steer their efforts towards application scenarios in the direction of acoustic biodiversity monitoring. As already said, the development of such scalable methods and technology is essential for the implementation of scalable biodiversity monitoring and assessment studies. In the following, we briefly mention in chronological order some technology evaluation campaigns organized or supported by the SABIOD project[56] because these considered large number of species, the largest among all at that time.

1. *The ICML 2013 Bird Challenge*[57,58] required an acoustic recognition/clustering of 35 bird species in continuous recordings. There were 35 training files and 90 test files in total provided by the National Museum of Natural

---

[51] IEEE International Conference on Acoustics, Speech and Signal Processing (ICASSP), http://2017.ieeeicassp.org/
[52] uLearnBio: Workshop on Unsupervised Learning from Bioacoustic Big Data,
[53] ICML: The International Conference on Machine Learning, http://www.machinelearning.org/icml.html
[54] Annual IEEE International Workshop on Machine Learning for Signal Processing, http://mlsp2016.conwiz.dk/home.htm
[55] Neural Information Processing Systems Conference, https://papers.nips.cc/
[56] The SABIOD project: Scaled Acoustic BIODiversity platform, http://sabiod.univ-tln.fr/
[57] The ICML 2013 Bird Challenge, https://www.kaggle.com/c/the-icml-2013-bird-challenge
[58] The audio datasets of the ICML 2013 Bird Challenge, http://sabiod.univ-tln.fr/icml2013/BIRD_SAMPLES/

History (MNHN)[59] in Paris. The results of this technology evaluation campaign were reported at the *Workshop on Machine Learning for Bioacoustics* (Glotin et al. 2013b), which was held jointly with the International Conference on Machine Learning (ICML-2013). The organizers recorded all workshop presentations and provided online links to the videos (cf. page 5 in vol. 1 of the workshop proceedings) and the presentation slides[60] in vol. 2 (Glotin et al. 2013b).

2. *The NIPS 2013 Multi-label Bird Species Classification task*[61,62] invited research teams to participate in acoustic recognition of 87 bird species, present into 1,000 continuous field recordings. The audio recordings were provided by the BIOTOPE society.[63] Seventy-seven teams participated in this technology evaluation campaign. The competition results were presented at the workshop *Neural Information Processing Scaled for Bioacoustics: from Neurons to Big Data* (Glotin et al. 2013c), held jointly with the NIPS-2013 conference.

3. *The LifeCLEF 2014 Bird Task*[64] was the first to bring the problem of scalability of methods and technology to the attention of the bioacoustic community. The organizers provided over 14,000 audio recordings from the *xeno-canto*[65] archive, which are representative for 500 bird species from Brazil. Each species is represented with multiple field recordings made by ten or more different persons, in different location and time. This challenge required the detection of the top-*k* species in each recording, with and without considering the background sounds. The recognition performance was measured separately for these two cases, also with and without the use of metadata. In total, ten teams participated in this challenge. The BirdCLEF2014 data sets and the recognition results are available at the LifeCLEF 2014 website[64] and a description of methods in (Goeau et al. 2014).

4. *The LifeCLEF 2015 Bird task*[66] increased the difficulty of large-scale bird recognition problem by considering 999 bird species from the *xeno-canto*[65] archive. The data set consisted of over 33,200 recordings made in Brazil, Colombia,

---

59 The National Museum of Natural History (MNHN) in Paris, https://www.mnhn.fr/en
60 The ICML 2013 Bird Challenge presentation slides, http://sabiod.org/ICML4B2013_proceedings_slides.pdf
61 The Neural Information Processing Scaled for Bioacoustics (NIPS4B), http://sabiod.univ-tln.fr/nips4b/challenge1.html
62 The Multi-label Bird Species Classification task, NIPS 2013, https://www.kaggle.com/c/multilabel-bird-species-classification-nips2013
63 The BIOTOPE society, http://www.biotope.fr/
64 The LifeCLEF 2014 Bird Task, http://www.imageclef.org/2014/lifeclef/bird
65 The *xeno-canto* archive, http://www.xeno-canto.org/
66 The LifeCLEF 2015 Bird task, http://www.imageclef.org/lifeclef/2015/bird

Venezuela, Guyana, Suriname, and French Guiana. Each bird species was represented by 14 to over 200 recordings, and the number of recordists per species ranged between 10 and 40. The data sets and the results for the six participants that submitted results are available at the LifeCLEF 2015 website[66] and description of methods in Joly et al. (2015).

5. The LifeCLEF 2016 Bird task[67] added a new dimension of difficulty by introducing species recognition in soundscape recordings. This was a major novelty with respect to the bird recognition competitions in the previous years. In fact, previous evaluation set-ups guaranteed that in each audio file there is one distinguishable dominant (foreground) species, which meant its sounds events are high above the acoustic background. The data set size of LifeCLEF 2016 was alike the LifeCLEF 2015 except that many of the audio files were soundscape recordings that contain multiple species with loud sound events. Another novelty was that for the first time participants were allowed to use additional training data sets if they wish or possess such data. The BirdCLEF 2016 data sets and the results for the six participants that submitted results are available online at the LifeCLEF 2016 website[67] and description of methods in the conference proceedings of CLEF-2016.

The LifeCLEF technology evaluation campaigns were co-organized by the SABIOD project[56], *xeno-canto*[65] archive, and the Pl@ntNet project.[68] Other smaller-scale technology evaluation challenges, such as

1. the MLSP 2013 Bird Classification Challenge,[69] which considered only 19 bird species represented in 645 files,
2. the EADM Challenge 2015,[70] which considers 11 categories of soundscapes,

also attracted some attention in the bioacoustic research community as these consider set-ups somehow relevant to the long-term goal – the development of methods and technology for large-scale biodiversity assessment and monitoring.

The interested reader might refer to the LifeCLEF[71] website and the Kaggle[72] platform for up-to-date information about the latest technology evaluation challenges.

---

[67] The LifeCLEF 2016 Bird task, http://www.imageclef.org/lifeclef/2016/bird
[68] The Pl@ntNet project, http://www.plantnet-project.org/papyrus.php?langue=en
[69] The MLSP 2013 Bird Classification Challenge, https://www.kaggle.com/c/mlsp-2013-birds
[70] The EADM Challenge 2015, http://sabiod.univ-tln.fr/eadm/#challenge
[71] The Home Page of the ImageCLEF / LifeCLEF, http://www.imageclef.org/
[72] The Kaggle website, https://www.kaggle.com/competitions

Concluding remarks

In response to the increased demand for automation support to biodiversity assessment and monitoring studies, the growing *computational bioacoustics* community established efficient mechanisms for the coordination of efforts, consolidation of resources, and periodic evaluation of advances in methods and technology. Collectively, these provided significant support to technology development, helped for steering and speeding up the evolution of methods and technology, and provided excellent collaboration opportunities. In this regard, university laboratories, governmental institutions, non-governmental organizations, research institutes, and other publicly or privately funded organizations started to share resources and efforts in order to research and develop robust species recognition methods. Last but not least, crowdfunded initiatives were launched in support of activities aiming at the development of smart tools, applications, and services in support of research tasks within the scope of computational bioacoustics. These are small yet important contributions to the efforts to provide scalable technological support to biodiversity assessment and monitoring studies.

The LifeCLEF technology evaluation campaigns, which are held annually since the year 2014, brought forward the important problem of robustness, operation on raw soundscape recordings, and scalability of methods and technology. As we already emphasized throughout the exposition in previous chapters, the availability of scalable methods and robust technology is a prerequisite for the creation of advanced technological tools and services, which would eventually provide efficient support to the implementation of the much-needed large-scale, long-term, and continuous biodiversity monitoring studies.

Epilogue

The emergence of tools and services developed by computational bioacoustic research projects gradually creates the ground for the implementation of large-scale, long-term, and continuous data collection and analysis studies. Such studies are attractive because they
- advance our understanding about the role of animal sound emissions,
- allow in-depth investigation of the importance of sound communication for the survival of certain species, and
- provide data for the assessment of influence of noise pollution and other energy pollutions to the health of entire ecosystems.

In such a way, computational bioacoustics undoubtedly contributes to the advances in environmental science. The advance of environmental science and the improved public awareness about the consequences of biodiversity loss, which translate to impoverishment of environment and lessening its capacity to support humankind, bring the issue of prioritizing sustainable economic activities to the political agenda.

Eventually, governments across the globe set targets for reducing the speed of biodiversity loss and already launched initiatives in support of some corrective polices.[1,2,3,4] The considerable rise of public attention towards the accelerating loss of biodiversity and the growing governmental support towards conservation planning and environmental impact assessment of human economic activities (Mandelik et al. 2005, Rands et al. 2010, Collen et al. 2013) brought forward the necessity of implementing rapid biodiversity surveys and long-term biodiversity monitoring programmes. All this raises an explicit demand of intelligent tools, services, and automated technology that could facilitate scalable biodiversity

---

[1] The most influential financial instruments created in the European Union in support of these polices is the LIFE Programme (http://ec.europa.eu/environment/life/) and its successor the LIFE+ Programme. (LIFE is the EU's financial instrument supporting environmental, nature conservation, and climate action projects throughout the EU. According to official information of the LIFE unit since 1992, LIFE has co-financed over 4,150 projects, contributing approximately 3.4 billion euros to the protection of the environment and climate.)

[2] Another instrument in support of biodiversity protection is the BiodivERsA network of 31 research-funding agencies across 18 European countries (http://www.biodiversa.org/). It is an ERA-NEt Co-fund, funded under the EU's Horizon 2020 Framework Programme for Research and Innovation.

[3] The National Science Foundation (NSF) of the USA, http://www.nsf.gov/

[4] The most influential instrument in Brazil is the National Council for Scientific and Technological Development (CNPq), http://cnpq.br/

DOI 10.1515/9781614516316-010

monitoring and environmental impact assessment studies. It is computational bioacoustics together with related disciplines such as acoustic ecology and soundscape ecology that possesses the potential to respond to this challenge and provide some of the essential means for achieving the objectives of scalable biodiversity monitoring and environmental impact assessment.

Furthermore, it is quite likely that the advances of artificial intelligence will make possible the creation of large machine minds, with data storage and processing capacity significantly superior to human brain. Once such a large machine mind is loaded up with all available knowledge from biology, ecology, acoustic ecology, soundscape ecology, landscape ecology, urban and environmental acoustics, behavioural ecology, biosemiotics, and other relevant disciplines, it will be able to provide multiple points of view to the same problem and suggest an acceptable trade-off between multiple (conflicting) objectives. We believe that making use of such combined knowledge, and methods, tools, and services for information acquisition and retrieval developed in computational bioacoustics, large machine minds will enable the automation of scalable biodiversity monitoring and environmental impact assessment studies.

Besides, we deem that the availability of large machine minds will eventually eliminate the present-day fragmentation of scientific disciplines and the current trend of differentiating countless small research areas, each supported by a small community of highly specialized researchers. Such a multifaceted approach to science is quite convenient to modern society as it facilitates the management of practical issues related to the limitations of humans with respect to brain capacity, training time, career development, the administration of funding bodies and projects, institutional organization and profiling, and so on. However, such a fragmentation does not help too much when complex multidisciplinary problems have to be addressed. In fact, this fragmentation creates obstacles and makes more difficult the resolution of complex practical problems. This is because cross-disciplinary research requires significant allocation of resources and large timescales to evolve successful and deliver back to society.

Therefore, we deem that it will be quite natural for machine minds to facilitate the convergence of earth-related scientific disciplines to one living earth science, which pools together the knowledge and methods of artificial intelligence, biology, ecology, acoustics, behavioural ecology, biosemiotics, and so on, with contemporary earth science in order to model practical problems at different levels of abstraction. Such convergence will allow the machine mind to work with the full body of information, fusing knowledge of different sources, creating and assessing models from different points of view, filling missing data and bridging understanding gaps, resolving conflicts between various information streams, conflicting objectives, and so on.

However, until such a large machine mind technology becomes available we all will depend on our skills to participate in multidisciplinary collaborations with open minds and open hearts. Fortunately, computational bioacoustics is an excellent example of how all this can succeed. The many success stories of computational bioacoustics (including the few outlined in this book) provide good support to such beliefs.

References

Aide, T. M., Corrada-Bravo, C., Campos-Cerqueira, M., Milan, C., Vega, G., Alvarez, R. (2013). Real-time bioacoustics monitoring and automated species identification. *PeerJ*, vol. 1, p. e103. DOI: 10.7717/peerj.103

Alkoot, F. M., Kittler, J. (1999). Experimental evaluation of expert fusion strategies. *Pattern Recognition Letters*, vol. 20, no. 11, pp. 11–13.

Anderson, S. E., Dave, A. S., Margoliash, D. (1996). Template-based automatic recognition of birdsong syllables from continuous recordings. *The Journal of the Acoustical Society of America*, vol. 100, no. 2, August 1996, pp. 1209–1219.

Aristotle (350 BC). *History of Animals*. Volumes 1–9.

Au, W. W. L., Hastings, M. C. (Eds.) (2008). *Principles of Marine Bioacoustics. Series Modern Acoustics and Signal Processing*. Springer, New York, ISBN: 978-0387783642.

Baker, E., Price, B. W., Rycroft, S. D., Hill, J., Smith, V. S. (2015). BioAcoustica: A free and open repository and analysis platform for bioacoustics. Database: bav054. ISSN: 1758-0463. DOI: 10.1093/database/bav054

Bardeli, R. (2008). Algorithmic Analysis of Complex Audio Scenes, Ph.D dissertation, Universität Bonn, Department of Computer Science III, Bonn, Germany.

Bardeli, R. (2009). Similarity search in animal sound databases. *IEEE Transactions on Multimedia*, vol. 11, no. 1, pp. 68–76. DOI: 10.1109/TMM.2008.2008920

Bardeli, R., Wolff, D., Kurth, F., Koch, M., Tauchert, K.-H., Frommolt K.-H. (2010). Detecting bird sounds in a complex acoustic environment and application to bioacoustic monitoring. *Pattern Recognition Letters*, vol. 31, no. 12, pp. 1524–1534. DOI: 10.1016/j.patrec.2009.09.014

Baum, L. E. (1972). An inequality and associated maximization technique in statistical estimation for probabilistic functions of Markov processes. *Inequalities*, vol. 3, pp. 1–8.

Baum, L. E., Petrie, T. (1966). Statistical inference for probabilistic functions of finite state Markov chains. *Annals of Mathematical Statistics*, vol. 37, pp. 1554–1563.

Bayes, T. (1763). An Essay towards solving a problem in the doctrine of chances. *Philosophical Transactions of the Royal Society of London*, vol. 53, pp. 370–418.

Bengio, S., Mariethoz. J. (2001). Learning the decision function for speaker verification. Technical Report. IDIAP Research Report 00-40, IDIAP, January 2001.

Bengio, Y., Frasconi, P. (1996). Input-output HMM's for sequence processing. *IEEE Transactions on Neural Networks*. September 1996, vol. 7, no. 5, pp. 1231–1249.

Bishop, C. (2006). *Pattern Recognition and Machine Learning*. Springer, Singapore. 2006.

Blumstein, D. T., Mennill, D. J., Clemins, P., Girod, L., Yao, K., Patricelli, G., Deppe J. L., Krakauer, A. H., Clark, Ch., Cortopassi, K. A., Hanser, S. F., McCowan, B., Ali, A. M., Kirschel A. N. G. (2011). Acoustic monitoring in terrestrial environments using microphone arrays: applications, technological considerations and prospectus. *Journal of Applied Ecology*, vol. 48, pp. 758–767. DOI: 10.1111/j.1365-2664.2011.01993.x

Boswall, J., Couzens, D. (1982). Fifty years of bird sound publication in North America: 1931–1981. *Discography, American Birds*, vol. 36, no. 6, pp. 924–943. Available online: https://sora.unm.edu/sites/default/files/journals/nab/v036n06/p00924-p00943.pdf. Last accessed: July 30, 2016.

Bourlard, H. A., Morgan, N. (1994). *Connectionist Speech Recognition a Hybrid Approach*. Kluwer Academic Publishers, New York, ISBN: 0-7923-9396-1, 1994.

Bovik, A. (2005). *Handbook of Image and Video Processing* (2nd edition). Burlington: Elsevier Academic Press.

Brandes, T. S., Naskrecki, P., Figueroa, H. K. (2006). Using image processing to detect and classify narrow-band cricket and frog calls. *Journal of the Acoustical Society of America*, vol. 120, pp. 2950–2957.

Breiman, L. (1996). Bagging predictors. *Machine Learning*, vol. 24, pp. 123–140.

Breiman, L. (2001). Random forests. *Machine Learning*, vol. 45, no. 1, pp. 5–32.

Briggs, F. (2013). Multi-instance multi-label learning: algorithms and applications to bird bioacoustics. Ph.D. thesis, Oregon State University, November 25, 2013.

Briggs, F., Fern, X., Irvine, J. (2013a). Multi-label classifier chains for bird sound. *Proceedings of the ICML Workshop on Machine Learning for Bioacoustics*, 2013.

Briggs F., Huang,Y.-H., Raich, R., Eftaxias, K., Lei, Z., Cukierski, W. (2013b). The 9th annual MLSP competition: New methods for acoustic classification of multiple simultaneous bird species in a Noisy environment. *Proceedings of the 2013 IEEE International Workshop on Machine Learning for Signal Processing*, September 22–25, 2013, Southampton.

Briggs, F., Lakshminarayanan, B., Neal, L., Fern, X. Z., Raich, R., Hadley, S. J. K., Hadley, A. S., Betts, M. G. (2012). Acoustic classification of multiple simultaneous bird species: A multi-instance multi-label approach. *The Journal of the Acoustical Society of America*, vol. 131, no. 6, pp. 4640–4650.

Broomhead, D. S., Lowe, D. (1988). Multivariable functional interpolation and adaptive networks. *Complex Systems*, vol. 2, pp. 322–355.

Cadore, J., Gallardo-Antoln, A., Pelez-Moreno, C. (2011). Morphological processing of spectrograms for speech enhancement. *Advances in Nonlinear Speech Processing*, vol. LNAI-7015, pp. 224–231. DOI: 10.1007/978-3-642-25020-0_29

Chesmore E. D. (2001). Application of time domain signal coding and artificial neural networks to passive acoustical identification of animals. *Applied Acoustics*, vol. 62, pp. 1359–1374.

Chesmore E. D., Nellenbach C. (2001). Acoustic methods for the automated detection and identification of insects. *Acta Horticulturae*, vol. 562, pp. 223–231.

Chesmore, D. (2004). Automated bioacoustic identification of species. *Anais da Academia Brasileira de Ciências*, vol. 76, no. 2, pp. 435–440.

Chu, W., Blumstein, D. T. (2011). Noise robust bird song detection using syllable pattern-based hidden Markov models. *IEEE International Conference on Acoustics, Speech and Signal Processing, ICASSP-2011*, pp. 345–348. DOI: 10.1109/ICASSP.2011.5946411

CLEF-2016 (2016). LifeCLEF-2016: Multimedia life species identification challenges. September 2016, Évora, Portugal, http://clef2016.clef-initiative.eu/. Last accessed: July 30, 2016.

Clemins, P. J., Johnson, M. T. (2006). Generalized perceptual linear prediction features for animal vocalization analysis. *The Journal of the Acoustical Society of America*, vol. 120, pp. 527–534.

Clemins, P. J., Trawicki, M. B., Adi, K., Jidong Tao, Johnson, M. T. (2006). Generalized perceptual features for vocalization analysis across multiple species. *Proceedings of the IEEE International Conference on Acoustics, Speech and Signal Processing, ICASSP 2006*, May 14–19, 2006, vol. 1, pp. 253–256.

Collen, B., Pettorelli, N., Baillie J. E. M., Durant S. M. (2013). *Biodiversity Monitoring and Conservation: Bridging the Gap between Global Commitment and Local Action*. John Wiley & Sons, Ltd, ISBN: 978-1-4443-3292-6, DOI: 10.1002/9781118490747

Collier, T. C., Kirschel, A. N. G., Taylor, C. E. (2010). Acoustic localization of antbirds in a Mexican rainforest using a wireless sensor network. *The Journal of the Acoustical Society of America*, vol. 128, pp. 182–189.

Cooley, J. W., Lewis, P. A. W., Welch, P. D. (1967). Historical notes on the fast Fourier transform. *IEEE Transactions on Audio and Electroacoustics*, vol. 15, no. 2, pp. 76–79. DOI: 10.1109/tau.1967.1161903

Cooley, J. W., Tukey, J. W. (1965). An algorithm for the machine calculation of complex Fourier series. *Mathematics of Computation*, vol. 19, pp. 297–301. DOI: 10.2307/2003354

Cover, T., Hart, P. (1967). Nearest neighbor pattern classification. *IEEE Transactions on Information Theory*, vol. 13, pp. 21–27.

Crandall, I. B. (1917). The composition of speech. *Physical Review*, vol. 10, no. 1, July 1917, pp. 74–76.

Dempster, A. P., Laird, N. M., Rubin, D. B. (1977). Maximum likelihood from incomplete data via the EM algorithm. *Journal of the Royal Statistical Society, Series B*, vol. 39, no. 1, pp. 1–38.

Depraetere, M., Pavoine, S., Jiguet, F., Gasc, A., Duvail, S., Sueur, J. (2012). Monitoring animal diversity using acoustic indices: Implementation in a temperate woodland. *Ecological Indicators*, vol. 13, pp. 46–54.

Dietterich, T. (1998). An experimental comparison of three methods for constructing ensembles of decision trees: bagging, boosting, and randomization. *Machine Learning*, vol. 40, no. 2, pp. 139–157.

Dietterich, T. (2000). Ensemble methods in machine learning. In Kittler, J., Rolli, F. (eds.) *Multiple Classifier Systems*, Springer-Verlag, London, UK, pp. 1–15.

Eldridge, A. C., Casey, M., Moscoso, P., Peck M. (2016). A new method for ecoacoustics? Toward the extraction and evaluation of ecologically-meaningful soundscape components using sparse coding methods. *PeerJ*, vol. 4, p. e2108. DOI: 10.7717/peerj.2108

Elman, J. L. (1990). Finding structure in time. *Cognitive Science*, vol. 14, pp. 179–211.

Evan, W., Mellinger, D. (1999). Monitoring grassland birds in nocturnal migration. *Studies in Avian Biology*, vol. 19, pp. 219–229.

Eyben, F., Weninger, F., Gross, F., Schuller, B. (2013). Recent developments in openSMILE, the Munich open-source multimedia feature extractor. *Proceedings of the ACM Multimedia (MM)*, Barcelona, Spain, October 2013, ACM, ISBN: 978-1-4503-2404-5. pp. 835–838. DOI: 10.1145/2502081.2502224

Farina, A. (2014). *Soundscape Ecology: Principles, Patterns, Methods and Applications*. ISBN: 978-94-007-7373-8, Springer Science+Business Media, Dordrecht, DOI: 10.1007/978-94-007-7374-5

Fisher, R. A. (1936). The use of multiple measurements in taxonomic problems. *Annals of Eugenics*, vol. 7, pp. 179–188.

Fodor, G. (2013). The ninth annual MLSP competition: first place. *Proceedings of the 2013 IEEE International Workshop on Machine Learning for Signal Processing (MLSP-2013)*, September 22–25, 2013, pp. 1–2. DOI: 10.1109/MLSP.2013.6661932

Frisk, G., et al. (2003). *Ocean Noise and Marine Mammals, National Research Council of the National Academies*. The National Academies Press, Washington DC. Available online at: http://www.nap.edu/read/10564/chapter/1. Last accessed: July 30, 2016.

Frommolt, K.-H., Bardeli, R., Clausen M. (Eds.) (2008). Computational bioacoustics for assessing biodiversity. *Proceedings of the International Expert Meeting on IT-based Detection of Bioacoustical Patterns*, December 7–10, 2007 at the International Academy for Nature Conservation (INA), Isle of Vilm, Germany. BfN – Skripten 234, 2008.

Frommolt, K.-H., Bardeli, R., Kurth, F., Clausen, M. (2006). The animal sound archive at the Humboldt-University of Berlin: Current activities in conservation and improving access for bioacoustic research. Slovenska akademija znanosti in umetnosti. Available online: https://books.google.bg/books?id=4DQSOgAACAAJ. Last accessed: July 30, 2016.

Gabor, D. (1946). Theory of communication. *Journal of the Institution of Electrical Engineers*, vol. 93, no. 26, pp. 429–457, DOI: 10.1049/ji-3-2.1946.0074

Ganchev, T. (2005). Speaker recognition. Ph.D thesis, Department of Electrical and Computer Engineering, University of Patras, Greece. Available online at: http://nemertes.lis.upatras.gr/jspui/bitstream/10889/308/3/PhDThesis.pdf . Last accessed: July 30, 2016.

Ganchev, T. (2009a). Enhanced training for the locally recurrent probabilistic neural networks. *International Journal of Artificial Intelligence Tools*, vol. 18, no. 6. pp. 853–881. ISSN: 0218-2130.

Ganchev, T. (2009b). Locally recurrent neural networks and their applications, In Soria, E., Martin, J. D., Magdalena, R., Martinez, M., Serrano, A. J. (eds.) Chapter IX in *Handbook of Research on Machine Learning Applications and Trends: Algorithms, Methods and Techniques*, ISBN: 978-1-60566-766-9, IGI Global, August 2009, pp. 195–222.

Ganchev, T. (2011). Contemporary methods for speech parameterization (1st edition). Springer, New York, 2011. ISBN: 978-1-4419-8446-3, e-ISBN: 978-1-4419-8447-0. DOI: 10.1007/978-1-4419-8447-0

Ganchev, T., Fakotakis, N., Kokkinakis, G. (2005). Comparative evaluation of various MFCC implementations on the speaker verification task. *Proceedings of the SPECOM-2005*, October 17–19, 2005. Patras, Greece, vol. 1, pp. 191–194.

Ganchev, T., Jahn, O., Marques, M. I., de Figueiredo, J. M., Schuchmann, K.-L. (2015). Automated acoustic detection of *Vanellus chilensis lampronotus*. *Expert Systems with Applications*, vol. 42, no. 15–16, pp. 6098–6111, ISSN: 0957-4174, DOI: 10.1016/j.eswa.2015.03.036

Ganchev, T., Parsopoulos, K. E., Vrahatis, M. N., Fakotakis, N. (2008). Partially connected locally recurrent probabilistic neural networks. In Hu, X., Balasubramaniam P. (eds.) *Recurrent Neural Networks* (Chapter 18), ARS Publishing, Vienna, Austria, pp. 377–400, ISBN: 978-3-902613-28-8.

Ganchev, T., Potamitis, I. (2007). Automatic acoustic identification of singing insects. *Bioacoustics: The International Journal of Animal Sound and its Recording*, vol. 16, no. 3. pp. 281–328, ISSN: 0952-4622.

Ganchev, T., Riede, K., Potamitis, I., Stoyanova, T., Ntalampiras, S., Dimitriou, V., Nomikos, B., Birkos, K., Jahn, O., Fakotakis N. (2011). AmiBio: Automatic acoustic monitoring and inventorying of biodiversity, Poster at the IBAC2011, *the XXIII meeting of the International Bioacoustics Council* (IBAC), in La Rochelle, France, September 12–16, 2011.

Ganchev, T., Tasoulis, D. K., Vrahatis, M. N., Fakotakis, N. (2007). Generalized locally recurrent probabilistic neural networks with application to text-independent speaker verification. *Neurocomputing*, vol. 70, no. 7–9, pp. 1424–1438.

Ganchev, T., Tasoulis, D. K., Vrahatis, M. N., Fakotakis, N. (2003). Locally recurrent probabilistic neural network for text-independent speaker verification. *Proceedings of the 8th European Conference on Speech Communication and Technology*, EUROSPEECH-2003, vol. 3, pp. 1673–1676.

Ganchev, T., Tsopanoglou, A., Fakotakis, N., Kokkinakis, G. (2002). Probabilistic neural networks combined with GMMs for speaker recognition over telephone channels. *Proceedings of the DSP2002*, Santorini, Greece, vol. 2, July 1–3, 2002, pp. 1081–1084.

Gasc, A., Sueur, J., Pavoine, S., Pellens, R., Grandcolas, P. (2013). Biodiversity sampling using a global acoustic approach: contrasting sites with microendemics in New Caledonia. *PLoS ONE*, vol. 8, no. 5, p. e65311.

Geurts, P., Ernst, D., Wehenkel, L. (2006). Extremely randomized trees. *Machine Learning*, vol. 63, no. 1, pp. 3–42.

Giannoulisy, D., Benetosx, E., Stowelly, D., Rossignolz, M., Lagrangez, M., Plumbley, M. D. (2013). IEEE AASP challenge detection and classification of acoustic scenes and events, hosted at WASPAA 2013. Technical Report, March 2013.

Gibaja, E., Ventura, S. (2010). Tutorial on multi-label learning. *ACM Computing Surveys*, vol. 9, no. 4, article 39, March 2010. DOI: 10.1145/2716262

Gibaja, E., Ventura, S. (2014). Multi-label learning: a review of the state of the art and ongoing research. *WIREs Data Mining Knowledge Discovery*, vol. 4, no. 6, pp. 411–444. DOI: 10.1002/widm.1139

Glotin, H., Sueur, J., Artières, T., Adam, O., Razik, J. (2013a). Sparse coding for scaled bioacoustics: From Humpback whale songs evolution to forest soundscape analyses. *The Journal of the Acoustical Society of America*, vol. 133, pp. 3311–3311. DOI: 10.1121/1.4805502

Glotin, H., Clark, C., LeCun, Y., Dugan, P., Halkias, X., Sueur, J. (Eds.) (2013b). *Proceedings of the First Workshop on Machine Learning for Bioacoustics*, joint to ICML-2013, Atlanta, ISSN: 979-10-90821-02-6, Available online at: http://sabiod.univ-tln.fr/ICML4B2013_book.pdf. Last accessed: July 30, 2016.

Glotin, H., LeCun, Y., Artières, T., Mallat, S., Tchernichovski, O., Halkias, X. (Eds.) (2013c). The NIPS4B 2013 competition. *Proceedings of Neural Information Processing Scaled for Bioacoustics: from Neurons to Big Data*, workshop joint to NIPS–2013, Nevada, ISSN: 979-10-90821-04-0. Available online at: http://sabiod.org/nips4b. Last accessed: July 30, 2016.

Goeau, H., Glotin, H., Vellinga, W.-P., Planque, R, Rauber, A., Joly, A. (2014). LifeCLEF Bird Identification Task 2014. CLEF: Conference and Labs of the Evaluation Forum, September 2014, Sheffeld, United Kingdom. 2014, Information Access Evaluation meets Multilinguality, Multimodality, and Interaction. <hal-01088829>. pp. 585–597. Available online at: http://clef2014.clef-initiative.eu. Last accessed: July 30, 2016.

Graciarena, M., Delplanche, M., Shriberg, E., Stolcke, A., Ferrer, L. (2010). Acoustic front-end optimization for bird species recognition. *Proceedings of the 2010 IEEE International Conference on Acoustics Speech and Signal Processing* (ICASSP-2010), March 14–19, 2010, pp. 293–296.

Greenwood, D. D. (1961). Critical bandwidth and the frequency coordinates of the basilar membrane. *The Journal of the Acoustical Society of America*, vol. 33, no. 10, pp. 1344–1356.

Greenwood, D. D. (1990). A cochlear frequency-position function for several species–29 years later, The Journal of the Acoustical Society of America, vol. 87, no. 6, pp. 2592–2605.

Greenwood, D. D. (1996). Comparing octaves, frequency ranges, and cochlear-map curvature across species. *Hearing Research*, vol. 94, no. 1–2, pp. 157–162.

Greenwood, D. D. (1997). The Mel Scale's disqualifying bias and a consistency of pitch-difference equisections in 1956 with equal cochlear distances and equal frequency ratios. *Hearing Research*, vol. 103, no. 1–2, pp. 199–224.

Halkias C. H., Paris S., Glotin H. (2013). Machine learning for Whale acoustic classification. *Journal of the Acoustical Society of America*, vol. 134, no. 5, pp. 3496–3505.

Hansen, L. K., Salamon, P. (1990). Neural network ensembles. *IEEE Transactions on Pattern Analysis and Machine Intelligence*, vol. 12, no. 10, pp. 993–1001.

Hansen, L. P. (1982). Large sample properties of generalized method of moments estimation. *Econometrica*, vol. 50, pp. 1029–1054.

Harris, F. J. (1978). On the use of windows for harmonic analysis with the discrete Fourier transform. *Proceedings of the IEEE*, vol. 6, pp. 51–83. DOI: 10.1109/PROC.1978.10837

Heijmans, H., Ronse, C. (1990). The algebraic basis of mathematical morphology. Part I: Dilations and erosions. *Computer Vision, Graphics and Image Processing*, vol. 50, pp. 245–295.

Henríquez, A., Alonso, J. B., Travieso, C. M., Rodríguez-Herrera, B., Bolaños, F., Alpízar, P., López-de-Ipina, K., Henríquez, P. (2014). An automatic acoustic bat identification system based on the audible spectrum. *Expert Systems with Applications*, vol. 41, no. 11, pp. 5451–5465. ISSN: 0957-4174, DOI: 10.1016/j.eswa.2014.02.021

Hill, D., Fasham, M., Tucker, G., Shewry, M., Shaw, P. (Eds). (2005). *Handbook of Biodiversity Methods: Survey, Evaluation and Monitoring*. Cambridge University Press, Cambridge, ISBN: 978-0-511-12535-5

Hinton, G. E., Sallans, B., Ghahramani, Z. (1998). A hierarchical community of experts. In Jordan, M. I. (ed.) *Learning in Graphical Models*, MIT Press, Cambridge, MA, USA, pp. 479–494, 1998.

Ho, T. K., Hull, J. J., Srihari, S. N. (1994). Decision combination in multiple classifier systems. *IEEE Transactions on Pattern Analysis and Machine Intelligence*, vol. 16, no. 1, pp. 66–75.

Huang, C.-J., Yang, Y.-J., Yang, D.-X., Chen, Y.-J. (2009). Frog classification using machine learning techniques. *Expert Systems with Applications*, vol. 36, pp. 3737–3743.

Huang, C.-J., Gao, W., Zhou, Z.-H. (2014). Fast multi-instance multi-label learning. *Proceedings of the 28th AAAI Conference on Artificial Intelligence*, pp. 1868–1874.

IEEE AASP Challenge detection and classification of acoustic scenes and events 2016, DCASE 2016 Workshop, Budapest, Hungary, September 2016, http://www.cs.tut.fi/sgn/arg/dcase2016/. Last accessed: July 30, 2016.

Jahn, O. (2011). Bird Communities of the Ecuadorian Chocó: A Case Study in Conservation. In Schuchmann, K.-L. (ed.) *Bonner Zoologische Monographien*, Nr. 56, ISBN: 978-3-925382-60-4, Zoological Research Museum A. Koenig (ZFMK), Bonn, Germany.

Jahn, O., Mporas, I., Potamitis, I., Kotinas, I., Tsimpouris, C., Dimitrou, V., Kocsis, O., Riede, K., Fakotakis N. (2013). The AmiBio Project – automating the acoustic monitoring of biodiversity, Poster at the IBAC2013, *the XXIV meeting of the International Bioacoustics Council*, Brazil.

Jahn, O., Riede, K. (2013). Further progress in computational bioacoustics needs open access and registry of sound files, IBAC-2013, Brazil.

Jebara, T. (2004). *Machine Learning: Discriminative and Generative*. Kluwer Academic Publishers, ISBN: 1-4020-7647-9.

Joly, A., Goeau, H., Glotin, H., Spampinato, C., Bonnet, P., Vellinga, W.-P., Planque, R., Rauber, A., Palazzo, S., Fisher, B., et al. (Eds). (2015). LifeCLEF 2015: Multimedia Life Species Identification Challenges. Working Notes of CLEF 2015 – Conference and Labs of the Evaluation forum – Toulouse, France, September 8–11, 2015. <hal-01182782>. Book Chapter in *Experimental IR Meets Multilinguality, Multimodality, and Interaction: 6th International Conference of the CLEF Association*, CLEF–2015, Springer International Publishing, July 2015. ISSN: 978-3-319-24027-5, DOI: 10.1007/978-3-319-24027-5_46, pp. 462–483

Jordan, M. I. (1986). Serial order: A parallel distributed processing approach. Institute for Cognitive Science Report 8604. University of California, San Diego.

Jordan, M. I., Jacobs, R. A. (1994). Hierarchical mixtures of experts and the EM algorithm. *Neural Computation*, vol. 6, no. 2, pp. 181–214.

Kaewtip, K., Tan, L. N., Alwan, A., Taylor, C. E. (2013). A robust automatic bird phrase classifier using dynamic time-warping with prominent region identification. *Proceedings of the 2013 IEEE International Conference on Acoustics, Speech and Signal Processing ICASSP-2013*, May 26–31, 2013, pp. 768–772. DOI: 10.1109/ICASSP.2013.6637752

Kinnunen, T., Li, H. (2009). An overview of text-independent speaker recognition: From features to supervectors. *Speech Communication*, ISSN: 0167-6393, vol. 52, no. 1, pp. 12–40. January 2010, DOI: 10.1016/j.specom.2009.08.009

Kittler, J., Alkoot, F. M. (2003). Sum versus vote fusion in multiple classifier systems. *IEEE Transactions on Pattern Analysis and Machine Intelligence*, vol. 25, no. 1, pp. 110–115, 2003.

Kittler, J., Hatef, M., Duin, R., Matas, J. (1998). On combining classifiers. *IEEE Transactions on Pattern Analysis and Machine Intelligence*, vol. 20, no. 3, pp. 226–239.

Kohonen, T. (1981). Automatic formation of topological maps of patterns in a self-organizing system. In Oja, E., Simula, O. (eds.) *Proceedings of the SCIA, Scand. Conference on Image Analysis*, Helsinki, Finland. pp. 214–220.

Kohonen, T. (1982). Self-organized formation of topologically correct feature maps. *Biological Cybernetics*, vol. 43, pp. 59–69.

Kohonen, T. (1995). Learning vector quantization, in Arbib, M. A. (eds.) *The Handbook of Brain Theory and Neural Networks*, MIT Press, Cambridge, MA, pp. 537–540.

Lam, L., Suen, C. Y. (1995). Optimal combinations of pattern classifiers. *Pattern Recognition Letters*, vol. 16, pp. 945–954.

Lang, K. J., Hinton, G. E. (1988). A time delay neural network architecture for speech recognition. Technical Report CMU-cs-88-152, Carnegie Mellon University, Pittsburgh PA.

Lasseck, M. (2013). Bird song classification in field recordings: Winning solution for NIPS4B 2013 competition. *Proceedings of the Workshop on Neural Information Processing Scaled for Bioacoustics*, joint to NIPS, Glotin H. et al. (ed.) http://sabiod.org/nips4b, Nevada, pp. 176–181.

Lasseck, M. (2014). Large-scale identification of birds in audio recordings: Notes on the winning solution of the LifeCLEF 2014 Bird Task, *Working Notes for CLEF 2014 Conference*, Sheffield, September 15–18, 2014. CEUR vol. 1180.

Lasseck, M. (2015). Improved automatic bird identification through decision tree based feature selection and bagging. *Working Notes of CLEF 2015 – Conference and Labs of the Evaluation forum*, Toulouse, France, Sept. 8–11, 2015. CEUR vol. 1391.

Lee, C.-H., Hsu, S.-B., Shih, J.-L., Chou C.-H. (2013). Continuous birdsong recognition using Gaussian mixture modeling of image shape features. *IEEE Transactions on Multimedia*, vol. 15, no. 2.

Lellouch, L., Pavoine, S., Jiguet, F., Glotin, H., Sueur, J. (2014). Monitoring temporal change of bird communities with dissimilarity acoustic indices. *Methods in Ecology and Evolution*, no. 4, pp. 495–505,

Leng, Y. R., Tran, H. D. (2014), MuItI-label bird classification using an ensemble classifier with simple features. *Proceedings of the 2014 Annual Summit and Conference Asia-Pacific Signal and Information Processing Association (APSIPA)*, pp. 9–12, December 2014, DOI: 10.1109/APSIPA.2014.7041649

Mandelik Y., Dayan T., Feitelson E. (2005). Planning for biodiversity: the role of ecological impact assessment. *Conservation Biology*, vol. 19, no. 4, pp. 1254–1261.

Mencía, E. L., Nam, J., Lee, D.-H. (2013). Learning multi-labeled bioacoustic samples with an unsupervised feature learning approach. *Proceedings of International Symposium Neural Information Scaled for Bioacoustics joint to NIPS*, Nevada, December 2013, Glotin H. et al. (eds.), ISSN: 979-10-90821-04-0.

Mennill, D. J., Vehrencamp, S. L. (2008). Animal duets are multifunctional signals: Evidence from microphone array recordings and multi-speaker playback. *Current Biology*, vol. 18, pp. 1314–1319.

Miller, D. C. (1916). *Science of the Musical Sounds*. Macmillan, New York, 1916.

Najman, L., Talbot, H. (Eds.) (2010). *Mathematical Morphology: From Theory to Applications*, ISTE-Wiley, ISBN: 978-1-84821-215-2.

Neal, L., Briggs, F., Raich, R., Fern, X. Z. (2011). Time-frequency segmentation of bird song in noisy acoustic environments. *Proceedings of the 2011 IEEE International Conference on Acoustics, Speech and Signal Processing*, ICASSP-2011, pp. 2012–2015. DOI: 10.1109/ICASSP.2011.5946906

Neto, J. Almeida, L., Hochberg, M., Martins, C., Nunes, L., Renals, S., Robinson, T. (1995). Speaker adaptation for hybrid HMM/ANN continuous speech recognition system. *Proceedings of the Eurospeech'95*, pp. 2171–2174.

NIST RT 2002–2009. NIST Rich Transcription (RT) series of challenges (2002–2009), http://www.nist.gov/itl/iad/mig/rt.cfm. Last accessed: July 30, 2016.

NIST SRE 1996–2012. NIST Speaker Recognition Evaluation (SRE) series of challenges (1996–2012), http://nist.gov/itl/iad/mig/sre.cfm. Last accessed: July 30, 2016.

Okun, O., Valentini, G., Re, M. (Eds.) (2011). *Ensembles in Machine Learning Applications*. Book Series: Studies in Computational Intelligence, vol. 373, Springer, ISBN: 978-3-642-22909-1.

Oliveira, A. G., Ventura, T. M., Ganchev, T. D., de Figueiredo, J. M., Jahn, O., Marques, M. I., Schuchmann, K.-L. (2015). Bird acoustic activity detection based on morphological filtering of the spectrogram. *Applied Acoustics*, vol. 98, pp. 34–42, ISSN: 0003-682X, DOI: 10.1016/j.apacoust.2015.04.014

Parsopoulos, K. E., Vrahatis, M. N. (2010). *Particle Swarm Optimization and Intelligence: Advances and Applications* (1st edition). Information Science Publishing (IGI Global), Hershey, PA, ISBN: 1615206663, DOI: 10.13140/2.1.3681.1206

Pavan G., Favaretto A., Bovelacci B., Scaravelli D., Macchio S., Glotin H. (2015). Bioacoustics and ecoacoustics applied to environmental monitoring and management, *Rivista Italiana di Acustica*, vol. 39, no. 2, pp. 68–74.

Pearlmutter, B. A. (1989). Learning state space trajectories in recurrent neural networks. *Proceedings of the International Joint Conference on Neural Networks IJCNN'89*, Washington DC, USA, June 18–22, 1989. Vol. 2, pp. 365–372.

Pearson, K. (1894). Contributions to the theory of mathematical evolution. *Philosophical Transaction of the Royal Society of London A*, vol. 185, pp. 71–110.

Pieretti, N., Farina, A., Morri, D. (2011). A new methodology to infer the singing activity of an avian community: the acoustic complexity index (ACI). *Ecological Indicators*, vol. 11, pp. 868–873.

Pijanowski, B. C, Villanueva-Rivera, L. J., Dumyahn, S. L., Farina, A., Krause, B. L., Napoletano, B. M., Gage, S. H., Pieretti, N. (2011a). Soundscape ecology: The science of sound in the landscape. *BioScience*, vol. 61, no. 3, pp. 203–216.

Pijanowski, B. C., Farina, A., Gage, S. H., Dumyahn, S. L., Krause, B. L. (2011b). What is soundscape ecology? An introduction and overview of an emerging new science. *Landscape Ecology*, vol. 26, no. 9, pp. 1213–1232. DOI: 10.1007/s10980-011-9600-8

Polak, E. (1997). *Optimization: Algorithms and Consistent Approximations*. Springer-Verlag. ISBN: 0-387-94971-2.

Popper A. N., Hawkins, A. (Eds.) (2012). *The Effects of Noise on Aquatic Life*. Series Advances in Experimental Medicine and Biology (Book 730), Springer, New York, ISBN: 978-1441973108.

Popper, A. N., Dooling, R. J. (2002). History of animal bioacoustics. *The Journal of the Acoustical Society of America*, vol. 112, pp. 2368–2368, DOI: 10.1121/1.4779607

Potamitis, I. (2014). Automatic classification of a taxon-rich community recorded in the wild. *PLoS ONE*, vol. 9, no. 5, p. e96936.

Potamitis, I., Ganchev, T. (2008). Generalized recognition of sound events: Approaches and applications. In Tsihrintzis, G. A., Jain, L. C. (eds.) Chapter 3 in *Multimedia Services in*

Intelligent Environments, Studies in Computational Intelligence (SCI) 120, Springer-Verlag, Berlin, Heidelberg, pp. 41–79, ISBN: 978-3-540-78491-3.

Potamitis, I., Ganchev, T., Kododimas, D. (2009). On automatic bioacoustic detection of pests: The cases of Rhynchophorus Ferrugineus and Sitophilus Oryzae. *Journal of Economic Entomology*, vol. 102, no. 4, pp. 1681–1690, ISSN: 0022-0493.

Potamitis, I., Ntalampiras, N., Jahn, O., Riede, K. (2014). Automatic bird sound detection in long real-world recordings: applications and tools. *Applied Acoustics*, vol. 80, no. 1–9. DOI: 10.1016/j.apacoust.2014.01.001

Potamitis, I., Rigakis, I. (2016a). Large aperture optoelectronic devices to record and time-stamp insects' wingbeats. *IEEE Sensors Journal*. vol. 16, no. 15, Aug. 1, 2016. pp. 6053–6061. DOI: 10.1109/JSEN.2016.2574762.

Potamitis, I., Rigakis, I. (2016b). Measuring the fundamental frequency and the harmonic properties of the wingbeat of a large number of mosquitoes in flight using 2D optoacoustic sensors. *Applied Acoustics*, vol. 109, pp. 54–60.

Potamitis, I., Rigakis, I., Fysarakis, K. (2014). The Electronic McPhail Trap. *Sensors*, vol. 14, no. 12, pp. 22285–22299.

Potamitis, I., Rigakis, I., Fysarakis, K. (2015). Insect biometrics: Optoacoustic signal processing and its applications to remote monitoring of McPhail type traps. *PLoS ONE* vol. 10, no. 11, p. e0140474. DOI:10.1371/journal.pone.0140474

Powell, M. J. D. (1987). Radial Basis Functions for multivariable interpolation: A review. In Mason, J., Cox, M. (eds.) *Algorithms for Approximation*. Clarendon Press, Oxford, pp. 143–167.

Pritchard, K., Eliot J. (2012). *Help the Helper: Building a Culture of Extreme Teamwork*, ISBN: 978-1-591-84545-4, Portfolio/Penguin, New York, USA.

Rands, M. R. W., Adams, W. M., Bennun, L., Butchart, S. H. M., Clements, A., Coomes, D., Entwistle, A., Hodge, I., Kapos, V., Scharlemann, J. P. W., Sutherland, W. J., Vira, B. (2010). Biodiversity conservation: Challenges beyond 2010. *Science*, vol. 329, no. 5997, pp. 1298–1303.

Ranft, R. (2004). Natural sound archives: Past, present and future. *Anais da Academia Brasileira de Ciencias*, vol. 76, no. 2, pp. 456–460. DOI: 10.1590/S0001-37652004000200041

Regen, J. (1903). Neue Beobachtungen über die Stridulationsorgane der saltatoren Orthopteren. *Arbeiten des Zoologischen Institut der Universität Wien*, vol. 14, no. 3, pp. 359–422.

Regen, J. (1908). Das tympanale Sinnesorgan von Thamnotrizon apterus Fab. als Gehörapparat experimentell nachgewiessen. Sitzungsberichte der Akademie Wissenschaften Wien. *Mathematisch-Naturwissenschaftlishe Klasse* (Abt. 1) III, vol. 117, pp. 487–490.

Regen, J. (1912). Experimentelle Untersuchungen uber das Gehor con Liogryllus campestris L., *Zoologisher Anzeiger*, vol. 40, no. 12, pp. 305–316.

Regen, J. (1913). Über die Anlockung des Weibchens von Gryllus campestris L. durch telephonisch übertragene Stridulationslaute des Männchens. *Pflüger's Archiv für die gesamte Physiologie des Menschen und der Tiere*, vol. 155, no. 1–2, pp. 193–200.

Regen, J. (1914). Untersuchungen über die Stridulation und das Gehör von Thamnotrizon apterus, Fab. Anzeiger Akademie der Wissenschaften. pp. 853–892. Retrieved August 28, 2014. Available online: http://www.landesmuseum.at/pdf_frei_remote/SBAWW_123_0853-0892.pdf. Last accessed: July 30, 2016.

Reynolds D. A. (1995). Speaker identification and verification using Gaussian mixture speaker models. *Speech Communication*, vol. 17, pp. 91–108.

Reynolds, D. A., Quatieri, T. F., Dunn, R. B. (2000). Speaker verification using adapted Gaussian mixture models. *Digital Signal Processing*, ISSN: 1051-2004, vol. 10, no. 1–3, pp. 19–41. DOI: 10.1006/dspr.1999.0361

Rosen, E. (2007). *The Culture of Collaboration: Maximizing Time, Talent, and Tools to Create Value in the Global Economy*, ISBN: 978-0-9774617-0-7, Red Ape Pub., 2007. Available online at Google Books: https://books.google.bg/books?id=glAsAQAAIAAJ. Last accessed: July 30, 2016.

Rosenblatt, F. (1958). The perceptron: a probabilistic model for information storage and organization in the brain. *Psychological Review*, vol. 65, pp. 386–408.

Rossing, Th. (Ed.) (2007). *Springer Handbook of Acoustics, Part A, Chapter 17: Animal Bioacoustics*, ISBN: 978-0-387-30446-5, Springer, pp. 473–490.

Saeidi, R., Mohammadi, H. R. S., Ganchev, T., Rodman, R. D. (2009). Particle swarm optimization for sorted adapted Gaussian mixture models. *IEEE Transactions on Audio, Speech and Language Processing*, vol. 17, no. 2, pp. 344–353. DOI: 10.1109/TASL.2008.2010278

Schmidt, M., Gish, H. (1996). Speaker identification via support vector classifiers. *Proceedings IEEE International Conference Acoustics, Speech, Signal Processing (ICASSP '96)*, Atlanta, vol. 1, pp. 105–108.

Schuchmann, K.-L., Marques, M. I., Jahn, O., Ganchev, T., Figueiredo, J. M. de (2014). Os Sons do Pantanal. CRBio-01. *O Biólogo*, vol. 29, pp. 12–15.

Setlur, A. R., Sukkar, R. A., Jacob, J. (1996). Correcting recognition errors via discriminative utterance verification. *Proceedings of ICSLP'96*, Philadelphia, vol. 2, pp. 602–605.

Sebastian-Gonzalez, E., Pang-Ching, J., Barbosa, J. M., Hart, P. (2015). Bioacoustics for species management: two case studies with a Hawaiian forest bird. *Ecology and Evolution*, vol. 5, no. 20, pp. 4696–4705.

Skowronski, M. D., Harris, J. G. (2004). Exploiting independent filter bandwidth of human factor cepstral coefficients in automatic speech recognition. *Journal of the Acoustical Society of America*, vol. 116, no. 3, pp. 1774–1780.

Soille, P. (1999). *Morphological Image Analysis: Principles and Applications* (2nd edition). Springer-Verlag, Berlin, Heidelberg GmbH, ISBN: 978-3-642-07696-1.

Somervuo, P., Kohonen, T. (1999). Self-organizing maps and learning vector quantization for feature sequences. *Neural Processing Letters*, vol. 10, no. 2, pp. 151–159.

Spampinato, C., Mezaris, V., van Ossenbruggen, J. (Eds). (2012). *Proceedings of the 1st ACM International Workshop on Multimedia Analysis for Ecological Data, ACM Multimedia Conference, MAED-2012*, Nara, Japan, October 29–November 02, 2012. New York. ACM. 433127.

Specht, D. F. (1966). Generation of polynomial discriminant functions for pattern recognition. Ph.D. Dissertation, Stanford University.

Specht, D. F. (1967). Generation of polynomial discriminant functions for pattern recognition. *IEEE Transactions on Electronic Computers*, vol. 16, pp. 308–319.

Specht, D. F. (1988). Probabilistic neural networks for classification, mapping, or associative memory. *Proceedings of IEEE Conference on Neural Networks*, San Diego. vol. 1, pp. 525–532.

Specht, D. F. (1990). Probabilistic neural networks. *Neural Networks*, vol. 3, no. 1, pp. 109–118.

Steen, K. A., Therkildsen, O. R., Karstoft, H., Green, O. (2012). A vocal-based analytical method for goose behaviour recognition. *Sensors*, vol. 12, pp. 3773–3788.

Storn, R., Price, K. (1997). Differential evolution – a simple and efficient adaptive scheme for global optimization over continuous spaces. *Journal of Global Optimization*, vol. 11, pp. 341–359.

Stowell, D., Giannoulis, D., Benetos, E., Lagrange, M., Plumbley, M. D. (2015). Detection and classification of audio scenes and events. *IEEE Transactions on Multimedia*, vol. 17, no. 10, 2015. pp. 1733–1746.

Stowell, D., Plumbley, M. D. (2014). Automatic large-scale classification of bird sounds is strongly improved by unsupervised feature learning. *PeerJ*, vol. 2, p. e488. DOI: 10.7717/peerj.488

Sueur, J., Farina, A., Gasc, A., Pieretti, N., Pavoine, S. (2014). Acoustic indices for biodiversity assessment and landscape investigation. *Acta Acustica united with Acustica*, vol. 100, no. 4, pp. 772–781.

Sueur, J., Pavoine, S., Hamerlynck, O., Duvail, S. (2008). Rapid acoustic survey for biodiversity appraisal. *PLoS ONE*, vol. 3, no. 12, p. e4065. DOI:10.1371/journal.pone.0004065

Thompson, D. W. (1910). *Historia Animalium*. The works of Aristotle. Volume 4. Ross, W. D., Smith, J. A. (eds.) Clarendon Press, Oxford. Available online: https://ia902504.us.archive.org/24/items/worksofaristotle04arisuoft/ worksofaristotle04arisuoft.pdf. Last accessed: July 30, 2016.

Todd, V. L. G., Todd, I. B., Gardiner J. C., Morrin E. (2015). *Marine Mammal Observer and Passive Acoustic Monitoring Handbook*, Pelagic Publishing. ISBN: 978-1907807664, http://www.marinemammalobserverhandbook.com/. Last accessed: July 30, 2016.

Towsey, M., Wimmer, J., Williamson, I., Roe, P. (2014). The use of acoustic indices to determine avian species richness in audio-recordings of the environment. *Ecological Informatics*, vol. 21, pp. 110–119.

Trifa, V., Girod, L., Collier, T., Blumstein, D. T., Taylor, C. E. (2007). Automated wildlife monitoring using self-configuring sensor networks deployed in natural habitats. *International Symposium on Artificial Life and Robotics* (AROB 2007), Beppu, Japan, 2007.

Trifa, V., Kirschel, A., Taylor, C. (2008a). Automated species recognition of Antbirds in a Mexican rainforest using hidden Markov models. *Journal of the Acoustical Society of America*, vol. 123, pp. 2424–2431.

Trifa, V., Kirschel, A., Yao, Y., Taylor, C., Girod, L. (2008b). From bird species to individual songs recognition: automated methods for localization and recognition in real habitats using wireless sensor networks. In Frommolt, K.-H., Bardeli R., Clausen, M. (eds.) *Proceedings of the International Expert Meeting on Computational Bioacoustics for Assessing Biodiversity*, 2008.

Trone, M., Balestriero, R., Glotin, H., David, B. E. (2014). All clicks are not created equally: Variations in very high-frequency acoustic DAQ of the Amazon River dolphin (*Inia geoffrensis*), *Journal of the Acoustical Society of America*, vol. 136, no. 4, pp. 2217–2217, DOI: 10.1121/1.4900047

Trone, M., Glotin, H., Balestriero, R., Bonnett, D. E. (2015). Enhanced feature extraction using the Morlet transform on 1 MHz recordings reveals the complex nature of Amazon River dolphin (*Inia geoffrensis*) clicks. *Journal of the Acoustical Society of America*, vol. 138, p. 1904. DOI: 10.1121/1.4933985

Vapnik, V. N. (2000). *The Nature of Statistical Learning Theory*. Second edition. Springer-Verlag, New York, ISBN: 0-387-98780-0.

Ventura, T. M., Oliveira, A. G., Ganchev, T., Figueiredo, J. M., Jahn, O., Marques, M. I., Schuchmann, K-L. (2015). Audio parameterization method for improved bird identification. *Expert Systems with Applications*, ISSN: 0957-4174, vol. 42, no. 22, pp. 8463–8471. DOI: 10.1016/j.eswa.2015.07.002

Wolpert, D. H. (1996). The lack of a prior distinctions between learning algorithms. *Neural Computation*, vol. 8, pp. 1341–1390.

Wolpert, D. H., Macready, W. G. (1997). No free lunch theorems for optimization. *IEEE Transactions on Evolutionary Computation*, vol. 1, no. 1, pp. 67–82.

Wrightson, K. (2000). An introduction to acoustic ecology, soundscape. *The Journal of Acoustic Ecology*, vol. 1, no. 1, pp. 10–13.

Xu, L., Krzyzak, A., Suen C. Y. (1992). Methods of combining multiple classifiers and their applications to handwriting recognition. *IEEE Transactions on Systems Man and Cybernetics*, vol. 22, no. 3, pp. 418–435.

Xu, X. Frank, E. (2004). Logistic regression and boosting for labeled bags of instances. In Dai, H., Srikant, R., Zhang, C. (eds.), *Lecture Notes in Artificial Intelligence*, vol. LNAI-3056, pp. 272–281. Springer, Berlin.

Yang, S., Zha, H., Hu, B. (2009). Dirichlet-Bernoulli alignment: A generative model for multi-class multi-label multi-instance corpora. *Advances in Neural Information Processing Systems*. 22. MIT Press, Cambridge, MA, pp. 2143–2150.

Zarnik, B. (1929). Zivot i rad Ivana Regena. Priroda (in Croatian), no. 1. pp. 1–7. Retrieved August 28, 2014. Available online: http://library.foi.hr/priroda/pregled.aspx?z= 80&sql=SDDDDC%28C929%28DDDDC-DDD-SSDDD4&u. Last accessed: July 30, 2016.

Zha, Z., Hua, X., Mei, T., Wang, J., Qi, G., Wang, Z. (2008). Joint multi-label multi-instance learning for image classification. *Proceedings of the IEEE Conference on Computer Vision and Pattern Recognition*, pp. 1–8.

Zhang, C., Ma, Y.-Q. (Eds.) (2012). *Ensemble Machine learning: Methods and Applications*. Springer, ISBN: 1441993258.

Zhang, M., Wang, Z. (2009). MIMLRBF: RBF neural networks for multi-instance multi-label learning, Neurocomputing, vol. 72, no. 16–18, pp. 3951–3956.

Zhang, M.-L. (2010). A k-nearest neighbor based multi-instance multi-label learning algorithm. *Proceedings of the 22nd International Conference on Tools with Artificial Intelligence*, ICTAI-2010, Arras, France, pp. 207–212.

Zhang, M.-L., Zhou, Z.-H. (2014). A review on multi-label learning algorithms. *IEEE Transactions on Knowledge and Data Engineering*, vol. 26, no. 8, pp. 1819–1837, DOI: 10.1109/TKDE.2013.39

Zhang, X., Yuan, Q., Zhao, S., Fan, W., Zheng, W., Wang, Z. (2010). Multi-label classification without the multi-label cost. *Proceedings of the SIAM International Conference on Data Mining*, pp. 778–789. SIAM.

Zhou, Z.-H. (2012). *Ensemble Methods: Foundations and Algorithms* (1st edition). Data Mining and Knowledge Discovery Series, Chapman & Hall, CRC, ISBN: 1439830037.

Zhou, Z.-H., Zhang, M.-L. (2007a). Multi-instance multi-label learning with application to scene classification. In Schölkopf, B., Platt, J. C., Hofmann, T. (eds.) *Proceedings of the Advances in Neural Information Processing Systems, NIPS-2006*. MIT Press, Vancouver, Canada, Cambridge, MA, pp. 1609–1616.

Zhou, Z.-H., Zhang, M.-L. (2007b). Solving multi-instance problems with classifier ensemble based on constructive clustering. *Knowledge and Information Systems*, vol. 11, no. 2, Springer, pp. 155–170.

Zhou, Z.-H., Zhang, M.-L., Huang, S.-J., Li, Y.-F. (2012). Multi-instance multi-label learning. *Artificial Intelligence*, vol. 176, no. 1, pp. 2291–2320.

Index

1-D audio parameterization 89
2-D audio parameterization 101

abundance assessment task 57
acoustic activity detector 46
acoustic libraries 64
acoustic species identification 39
Acoustics Ecology Institute 194
AmiBio project 170
Animal Sound Archive 193
ARBIMON Acoustics project 164
ARBIMON II platform 169
ARBIMON II project 167
attention value 124
audio chunk 9
audio parameterization 93
audio pre-processing 90
audio pre-screening process 73
Australian National Wildlife Collection Sound Archive 193

bag-of-instances 127, 154
BioAcoustica repository 192
Bioacousticians 25
Bioacoustics 3, 4, 25
Bioacoustics e-mailing list 198
Biodiversity 1
blob 127
bootstrap dataset 88
Borror Laboratory of Bioacoustics Sound Archive 191
British Library Sound Archive 191

clustering of sound events 34, 58
cohort models 67
Computational Bioacoustics 3, 4, 22
contemporary Bioacoustics 17, 19
continuous soundscape recordings 180

decision trees 138
dilation 107
discriminative classification methods 137
discriminatively trained HMM 140
DORSA archive 194

ensemble learning methods 143
ENTOMATIC project 182
erosion 107
expectation-maximization algorithm 146

Gaussian Mixture Models 138
generative classification methods 138
GLR PNN 140
GMM-LR/SVM 140
GMM-UBM 143
ground true labels 65

hand editing of snippets 74
Hidden Markov Models 138
HMM/MLP 140

instance 127, 154

Linear Discriminant Analysis 137
localization and tracking of individuals 33, 52
LRPNN 140

Macaulay Library archive 191
MFCC 120
morphological filtering 107
morphological filtering of the audio spectrogram 114
morphological image processing 107
multi-instance multi-label classification 154
multi-instance multi-label classifier 142
multi-label random decision trees 142
multi-label species identification 41
multi-label species identification task 32
Multi-layer Perceptron 138
multi-species diarization task 33, 49

Nature Sounds Society 194
non-discriminative classification methods 137

one-category recognition 45
one-category recognition task 32
one-species detection task 32
one-species detection task 37
one-species recognition task 33, 47

DOI 10.1515/9781614516316-012

opening morphological operator 115
operator dilation 117
operator erosion 117
optoacoustic identification of mosquito species 185

Pantanal Biomonitoring Project 178
PC LRPNN 140
PNN-RNN hybrids 140
points of interest identification 121
Polynomial Classifier 137
Probabilistic Neural Network 138

Radial Basis Function 140
random forest decision trees 142
Recurrent Neural Networks 137
REMOSIS project 184

SABIOD platform 181
SABIOD project 181
scikit-learn 143
signal segmentation 91
single model per category 148
single-instance single-label 142
single-label multi-class classifier 148

snippet selection process 74
sound event type recognition task 54
Sound Library of the Wildlife Sound Recording Society 192
sound-event type recognition task 33
soundscape 6
species abundance assessment 34
species identification tasks 32
spectrogram thresholding 105
statistical standardization 98
structure tensor 121
structuring element 107, 115
Support Vector Machines 138

temporal derivative 98
Tierstimmenarchiv at the Museum für Naturkunde in Berlin 193
Time-Delay Neural Networks 137

variable length segmentation 92

Western Soundscape Archive 193
whitening filter 129

xeno-canto repository 190

Printed in Great Britain
by Amazon

4f588e03-e6a2-48a7-bd92-98776c08b97cR01